SELLING SCIENCE *Revised Edition*

How the Press Covers Science and Technology

DOROTHY NELKIN

W. H. FREEMAN AND COMPANY
New York

Library of Congress Cataloging-in-Publication Data

Nelkin, Dorothy.
    Selling science : how the press covers science and technology /
Dorothy Nelkin.—rev. ed.
        p.   cm.
    Includes index.
    ISBN 0-7167-2595-9 (pbk.)
    1. Science news.   2. Communication in science.   3. Communication
of technical information.   I. Title.
Q225.N35   1995
070.4'495—dc20   94-39524
                                                               CIP

© 1987, 1995 by W. H. Freeman and Company

Printed in the United States of America

1 2 3 4 5 6 7 8 9 0 VB 9 9 8 7 6 5 4

CONTENTS

PREFACE

Science is a part of common culture, integrally tied to social practices, public policies, and political affairs. Frequent reports of outrageous science scandals and technological risks remind us not only of our dependence on the media for timely information about science and technology but also of the limits of what we can learn from the press. People are not receiving the critical and comprehensive reporting about science and technology so essential in contemporary society, where most decisions rest on technological expertise. Although we depend on the media for science news, there is little understanding of the relationships between scientists and journalists that lie behind the images of science conveyed in the press.

My interest in this topic developed while I was doing research on public attitudes toward science and technology and, in particular, on technological controversies. I was struck by the ubiquitous tendency of scientists, engineers, and physicians to condemn the media, to criticize the quality of science reporting, and to attribute negative or naïve public attitudes about science and technology to popular news coverage of science-related events. At the same time, science professionals are often unable to document their complaints, to specify what is wrong. I therefore began to explore the images of science and technology presented to the public through the media, and the characteristics of both journalism and science that contribute to shaping these images.

My research has focused on a range of sources, including national newspapers such as *The New York Times, The Christian Science Monitor, The Wall Street Journal,* and *The Washington Post;* local newspapers, drawing from *Newsbank,* a file of 100 local newspapers around the United States that are indexed by categories to cover local views of selected public issues; national news magazines such as *Time, Newsweek,* and *U.S. News and World Report;* and widely distributed specialized magazines, including women's, health, and business magazines. Excluded from my analysis are specialized magazines, such as *Science,* which are directed to scientists. I have included some material on television reporting, but excluded science-fiction programs, medical dramas, educational programs, and docudramas that present fictionalized images of science with relatively little concrete information.[1] People turn to newspapers, magazines, and the evening television news to learn about current science and technology events. Television talk shows and special reports are increasingly important as a source of information about science. So too are science museums. But the print media continue to be the primary source of news in this area of reporting.[2]

To analyze the coverage of science and technology in the popular media, I selected a spectrum of issues that have attracted extensive coverage, looking for the dominant themes and recurring metaphors that project an image of science and technology to the public. The style and content of science reporting has changed considerably in the seven years since the first edition of this book. These changes reflect several factors: The increased scale of science has raised questions of social priorities and research costs; the growing importance of research in human biology has raised concerns about ethical implications; the many reports of scientific fraud have increased public mistrust; and the continuing incidents of technological risk have turned individual events into generic problems. The coverage of science in the 1990s has also been influenced by the growing competition in the media: events are dramatized and public figures villainized in the endless quest for "news." Such changes call for a reanalysis of science–media relationships. Thus, in this revised edition, I updated the earlier material, providing current examples and emphasizing the changes since this book was first written.

Studies of public communication of science and technology remain scattered, but as the media become more central to contemporary society, such works have multiplied. Aside from the newsletters of the science writers' profession, the first journal especially devoted to this topic, *Public Understanding of Science,* was established in London in 1992.[3] I have tried to bring together, in this edition, those studies that bear on the complex relations between scientists and journalists as they influence the coverage of science in the press. But my investigation led well beyond this literature. For material on scientists' concern about their image in the press, I turned to science policy journals and to the professional journals of science, engineering, and medicine. Some of the information on public relations activities of scientists and industries was gathered directly from public relations firms; some from discussions with scientists, journalists, and public relations officers; some from participation in meetings on the subject.

To better understand the nature of science journalism I spoke extensively to many reporters and participated in meetings, press conferences, and informal discussions. As a member of the Council for the Advancement of Science Writing, I had the privilege of spending many hours with some of the leading American science journalists and of listening to their professional concerns. I also gathered material from the Media Resources Service of the Scientists' Institute for Public Information, which serves as a liaison between scientists and journalists. Those quotations in the text that are not specifically referenced are from informal discussions and interviews with journalists and scientists.

I benefited from transcripts of interviews with journalists conducted by Sharon Friedman, from a survey by Bruce Lewenstein, and from news files collected by colleagues who have done research on the topics covered in Chapters 2 through 4. These people include Jon Beckwith, Stephen Hilgartner, Susan Lindee, Gerry Markle, Mary Marlino, Elena Nightingale, and Chris Anne Raymond. Roald Hoffmann and Kenneth Wilson made available their clippings on their Nobel Prizes. David Perlman and other journalists gave me access to their files. I also gained insights from the discussions of the Twentieth Century Fund Task Force on the Communication of Risk, for which I wrote the background paper (*Science in the Streets,* New York: Priority Press, 1984).

I am particularly indebted to Stephen Hilgartner for research assistance and for creative guidance in analyzing metaphors and images. Jordana Pestrong ably assisted me in the research for the second edition. Several people provided invaluable criticism of drafts of this manuscript, including Barbara Culliton, Sharon Dunwoody, Daniel X. Freedman, Sharon Friedman, Rae Goodell, Fred Jerome, Alex Keynan, Marcel LaFollette, Elena Nightingale, David Perlman, Carol Rogers, David Rubin, Michael Schudson, and David Zimmerman. My editor, Jonathan Cobb, contributed far more than editing, providing ideas and a great deal of help in shaping both editions of the book. And, as usual, Mark Nelkin was a critical and supportive reader.

Finally, research expenses and time for research and writing were provided by a Guggenheim Fellowship, a year as a visiting scholar at the Russell Sage Foundation, and grants from the National Science Foundation (R11–8510076) and from the National Institutes of Health (1RO1 HG0047-01).

Dorothy Nelkin
New York, New York
June 1994

1

SCIENCE AND TECHNOLOGY IN THE MEDIA

In 1924, Edwin E. Slosson, editor of the first science writing syndicate in America, described his view of science journalism. "The public that we are trying to reach is in the cultural stage when three-headed cows, Siamese twins and bearded ladies draw the crowds to the side shows." That is why, he explained, science is usually reported in short paragraphs ending in "-est." "The fastest or the slowest, the hottest or the coldest, the biggest or the smallest, and in any case, the newest thing in the world."[1]

In some respects little has changed. In the 1990s research on embryo cloning, pregnant postmenopausal women, and genetically engineered pigs is drawing readers and selling magazines. And journalists play up the biggest collider, the newest techniques of bioengineering, the riskiest technologies. Indeed, the style of reporting has been remarkably consistent over time.

Today, news about science and technology is featured in front page articles—in stories about discoveries, news about health, and reviews of economic trends and business affairs. Media attention focuses on technology-related policy issues such as environmental quality and public health. Controversies—over biotechnology, AIDS therapies, the patenting of new life forms, and incidents of fraud—have become newsworthy events.

Increasingly, in the 1990s, science appears in the coverage of such global issues as climate change, environmental disasters and international economic affairs. And scandals—from the radiation experiments on human subjects during the Cold War to the falsification of data for research on alternative breast cancer therapies—concentrate interest on the problems of science and call attention to the importance of timely and informative reporting.

For most people, the reality of science is what they read in the press. They understand science less through direct experience or past education than through the filter of journalistic language and imagery. The media are their only contact with what is going on in rapidly changing scientific and technical fields, as well as a major source of information about the implications of these changes for their lives. Good reporting can enhance the public's ability to evaluate science policy issues and the individual's ability to make rational personal choices; poor reporting can mislead and disempower a public that is increasingly affected by science and technology and by decisions determined by technical expertise.

At the community level, people are continually confronted with choices that require some understanding of scientific evidence: whether to allow the construction of a toxic waste disposal dump in their neighborhood, or how to respond to a child with AIDS in their school. Similar choices must be made at the personal level: whether to use estrogen replacement therapy, whether to eat high-fiber cereals or to reduce consumption of coffee, or how to act upon the results of a prenatal genetic test. Information and understanding are necessary if people are to think critically about the decisions they must make in their everyday lives.

What, then, is conveyed about science and technology in the press? In 1966 Frank Carey, a writer for the Associated Press, was asked this question. He listed the following news items that had been reported by science writers over the previous 20 years: "the explosion of a nuclear device in Red China . . . the launching of a flying doghouse by the Russians . . . the birth of quintuplets in South America . . . the sex-lure chemical by which the female German cockroach calls her boyfriend . . . the heartbeat of Olga, the whale . . . and such hot potatoes as fluoridation, Rachel Carson versus the bad guys . . . the radioactive fallout from nuclear bomb tests."[2] Today, nearly thirty years later, science journalists continue to report on a remarkable range of subjects. Indeed, they sometimes call themselves the SMEERSHS: "We cover Science, Medicine, Energy, Environment, Research, and all sorts of other SHit."

Science appears in the coverage of dramatic crises, major discoveries, and the feats and foibles of science stars. The applications or implications of scientific knowledge, dramatic or unusual events, and technical disputes have the greatest media appeal. But science also appears in news articles on drugs, food additives, transplants and artificial organs, cancer, genetic diseases, and new bioengineered products. And technical information is integrated into the news of the day: From descriptions of artificial heart transplants we find out about human physiology; from stories on AIDS we read about epidemiology and immunology; from reports on technological hazards we read about research on toxic substances and radiation risk; and from news about earthquakes, we hear about geology and seismology.

But what do we actually learn about science and technology? Consider, for example, the history of news reporting on interferon, a protein manufactured in the body when a virus invades a cell. Interferon was discovered in 1953 as a natural therapeutic agent, a so-called "interfering protein" that inhibits infection. The possibility of isolating the protein raised hopes for developing a cancer cure, and this caught media attention at the time. However, the scarce supply of the agent limited progress and the subject faded from public view until 1975 when Mathilde Krim, a politically astute geneticist, organized a conference intended to publicize the potential of interferon and to win public support for research.[3] Three years later the American Cancer Society (ACS) financed clinical trials to test the effectiveness of the protein.

Krim's efforts and the interest of the ACS brought a deluge of media coverage. The scientific press qualified the promises of interferon research, indicating the tentative nature of existing studies, the high cost of isolating the protein, and its therapeutic limits. In the popular press, however, interferon was a "magic bullet," a miracle cure for everything from cancer to the common cold.

In 1980 Biogen, a biotechnology firm, developed a DNA clone for the protein. The ability to synthesize interferon opened the possibility of producing large quantities of the product at low cost. Uncritically accepting promotional information from a Biogen press conference, journalists welcomed this new technological development as still another miracle. "Like the genie in a fairy tale," the *Detroit Free Press* told its readers, "science came up with the key to the magic potion, a way to produce interferon in bulk."

Reader's Digest talked about a "wonder therapy," *Newsweek* about "cancer weapons" and "the making of a miracle drug." Business journalists focused on interferon as a profitable commodity, calling attention to the dramatic increase in stock prices of biotechnology firms. *Business Week* described efforts to synthesize the substance as a "race" to capture the market: "We have just passed the quarter mile pole and all the horses are in a bunch." *Time* wrote of the "gold mine for patients and for companies." The *Saturday Evening Post* claimed that "punters in Wall Street are already laying bets that interferon is a sure winner."

Throughout this period, *New York Times* science writer Harold Schmeck wrote cautious reports, suggesting that interferon was promising but that there was no definitive evidence of its effectiveness. And he reported on possible harmful effects, suggesting that "the seemingly ideal weapon" was less of a panacea than anticipated. In May 1980 he reported on interferon studies that "put cancer use in doubt," emphasizing the "modest, controversial, and even negative results of research." He observed that the promise of a scientific advance raised research money, but also false hopes. In response to this article, four scientists from the Sloan Kettering Institute for Cancer Research wrote a letter to the *New York Times,* expressing concern that such qualified reporting could undermine public support of interferon research.[4]

By 1982 other journalists began to report on the toxic side effects of interferon. Though aware of these effects since the mid-1970s, scientists had provided no public information for fear it would dampen popular enthusiasm and stall the interferon crusade. But the difficulties of using interferon as a therapeutic agent became public in 1982 when four patients treated with interferon in France died. Abruptly, the tone of reporting changed from exaggerated optimism to disillusionment: "From wonder drug to wall-flower." The drug was demoted from a magic bullet and disease fighter to a mere "research tool." Newspapers and magazine articles assessed the situation pessimistically: "Jury's out on interferon as a cancer cure"; "Studies cast doubt on cancer drug"; "It's a hard row to hoe." Research continued, but little more appeared in the press until a series of patent disputes turned media attention to the question of proprietary interests in commercially promising biotechnology products.

The popular accounts of interferon research demonstrate several striking features of science reporting that I will develop in subsequent chapters. First, imagery often replaces content. Little appeared in the press coverage of interferon about the actual nature of the research; instead, most articles appealed to public concerns about cancer and hopes for a cure for this dread disease. While interferon's short-term usefulness as a therapeutic agent was problematic, the research did yield significant scientific understanding of basic biological concepts, such as the control of gene expression in mammalian cells and the regulation of immunity, that in the long term have affected the practice of medicine. But readers following the interferon story learned little about such developments. This style of reporting has continued in the recent coverage of human genome research where, as we will see in later chapters, imagery often replaces content. Conveyed are simple, condensed, and attention-seeking impressions—a "blueprint" of life, a "Book of Man," a "medical crystal ball"—images devoid of real meaning and useful information.

Second, the press covered interferon research as a series of dramatic events. Readers were treated to hyperbole, to promotional coverage designed to raise their expectations and whet their interest. When predictions about interferon's curative powers failed to materialize, however, unqualified optimism in the press quickly shifted to the opposite extreme. Similarly, many scientific events—possible AIDS therapies, potential new sources of energy, the discovery of genetic markers for disease—evoke premature enthusiasm and optimistic expectations. But then comes disillusionment when promises fail.

A third feature of science journalism revealed by the interferon reports is the focus on scientific and technological competition. Scientists and the firms developing interferon were in a "race" for breakthroughs, for solutions. The gradual accumulation of information inherent to the research process was not considered news. Whether the goal is to discover a genetic marker or a new energy source, the media always focus on the competition in science, the race to be the first to get results.

Perhaps the most surprising feature illustrated by the interferon story is the role that scientists played in promoting interferon and in shaping the media coverage of the research. Far from being neutral sources of information, scientists themselves actively sought a favorable press, equating public interest with research support.

As research funds decline in the 1990s, scientists are increasingly using rhetorical strategies to attract attention. We read of chaos and quarks, big bangs and black holes, bucky balls and superstrings, master molecules and medical crystal balls. The assumption is that media interest will influence those who control the purse. "Scientists," said Neal Lane, director of the National Science Foundation in 1993, must "sell themselves to the public to ensure that science retains both public support and the funding that goes with it."[5]

These features of science journalism include some curious contradictions. There is more science news in the media every year.[6] Yet public understanding of science and technology is in many ways distorted. This is an age of science fantasy and scientistic cults. While scientific rationality is valued as the basis of our "knowledge society," science is invested with magic and mystique; we expect "magic bullets" and "miracle cures." People who demand sophisticated science-based medicine may also support the animal rights movement in its opposition to the experiments that allow the development of therapeutic techniques. While we welcome technology as the key to progress and the solution to problems, we are increasingly preoccupied with risk, fearing the very technologies we most depend upon.

A further irony lies in the ambivalence of scientists toward the press.[7] Scientists employ increasingly sophisticated public relations techniques to assure that their interests are represented with maximum media appeal. Their efforts to attract media attention have increased during the past decade. This reflects the growth of large-scale, costly research tied closely to applications, the changing relationship between scientists and their traditional sources of funding, and the accountability demanded by a public concerned about the social implications of science and technology and inclined to question the credibility of scientific and technical institutions. Yet, despite their growing interest in media coverage, scientists mistrust journalists and criticize the reporting about their fields. They complain about inaccurate, sensational, and biased reporting and fear that the press encourages antiscience attitudes. Ironically, as media interest in science increases, so too do the tensions between scientists and journalists, for along with media attention comes greater public scrutiny and pressures for regulation.

Journalists themselves often criticize the way science is presented in the press. However, they tend to blame their sources—scientists and technical institutions—for providing misleading or inadequate information. Mutually dependent, the communities of science and journalism are wary collaborators in the business of science communication. Science writer William Burrows described their uneasy relationship: "Scientists think that whatever they tell a reporter is bound to come out wrong. . . . Most ordinary reporters would practically cross the street to avoid running into an expert since they consider scientists to be unemotional, uncommunicative, unintelligible creatures who are apt to use differential equations and logarithms against them the way Yankee pitchers use inside fast balls and breaking curves."[8]

This book attempts to explain these ironies by exploring what is going on behind the scenes of science journalism. As we read our magazines and newspapers, what do we find out about science and technology? And what messages emerge from the selective news we receive? What characteristics of journalism affect the creation of science news? And how do the public relations efforts of scientists influence the coverage of science in the media?

The media are a diverse enterprise. Nationally distributed newspapers such as the *New York Times, Washington Post,* and *Wall Street Journal* are influential not only because of their large readership, but because they serve as a point of reference, a bulletin board for government officials, for journalists who write for other newspapers, and for television reporting.[9] In effect, they establish a tone and a set of standards for journalism. The myriad local and regional newspapers mostly belong to major chains such as Gannett or Knight-Ridder, each chain with circulations of over 3 million. Their staff reporting is mainly directed to local issues, and they rely on the wire services for national or specialized news. Science stories are picked up from Associated Press (AP) and United Press International (UPI) and edited to reflect local interests. The large circulation weekly newsmagazines such as *Time, Newsweek,* and *U.S. News and World Report,* specialized magazines such as *Business Week,* and science magazines such as *Discover* are important sources of information about science, having the time (because of their weekly publication schedule) and space to expand on science news.[10]

There are, of course, significant differences between local newspapers, which have few specialized reporters, and national papers, which employ a stable of experienced science writers who know the science terrain. Yet a surprising feature of science journalism is its homogeneity. While journalistic reports on science and technology vary from the 60-second sound bite to the investigative report, most articles on a given subject focus on the same issues, use the same sources, and interpret the material in similar terms.[11]

Journalists are bound by similar cultural biases and professional constraints. Sharing common assumptions about science and technology, their writing on scientific issues and events takes place within what sociologist Todd Gitlin calls a frame; that is, "a persistent pattern of cognition, interpretation and presentation, of selection, emphasis, and exclusion."[12] This frame organizes the world for journalists, helping them to process large amounts of information, to select what is news, and to present it in an efficient form. Their metaphors, descriptive devices, and catch phrases are expressions of this frame.[13]

The journalistic approach to science reporting, however, has varied over time.[14] The 1960s was a period of scientific and technological "breakthroughs" and "revolutions." Journalists covered the cosmic events of the space program and the dramatic discoveries in the physical sciences with wonder and elan. The frame changed in the late 1960s and the 1970s, when wonder gave way to concern about environmental and social risks. Journalists shifted their attention at this time from the conquests of science and technology to their consequences, from the celebration of progress to a more critical reflection about the problems brought about by technological change. In the 1980s the technological enthusiasm of the 1960s was born again, though tempered by the continued fear of risk. The idea of progress was resurrected as innovation, and the celebration of technology turned to high-tech hype. The hype continues in the 1990s, though focused more on the biological than the physical sciences. The Human Genome Project has replaced the space program as the "new frontier." The goal is the discovery, not of outer space, but inner space—an idea especially appealing in this New Age. Today, however, the economic costs of big science projects and advanced medical technology, together with the ethical implications of biological research, are an important part of science news. So too are incidents of fraud. And in a more environmentally aware society, the media are reflecting on the global implications of technological change. There is today more critical, more negative, reporting about science and technology in the press.

These cyclical trends are apparent in the metaphors journalists use to describe science and technology. Although experienced science writers are more self-conscious about language and more restrained than the general reporter ("We never use the word 'breakthrough' anymore," science writers tell me), most science reporting shares a style, an imagery, and a particular world view.

Metaphors, a prevalent and important vehicle of public communication in all areas, are especially important in science reporting. Explaining and popularizing unfamiliar, complex, and frequently technical material can often be done most effectively through analogy and imagery. But metaphors are more than an aid to explanation; they are, as literary critic Kenneth Burke commented, "strategies . . . designed to organize and command the army of one's thoughts and images and so to organize them."[15]

Similarly, linguists George Lakoff and Mark Johnson insist that a metaphor is not just a rhetorical flourish, but a basic property of language used to define experience and to evoke shared meanings. Metaphors affect the ways we perceive, think, and act, for they structure our understanding of events, convey emotions and attitudes, and allow us to construct elaborate concepts about public issues.[16]

By their choice of words and metaphors, journalists convey certain beliefs about the nature of science and technology, investing them with social meaning and shaping public conceptions of limits and possibilities. Was Chernobyl a "disaster" or an "event"? Is dioxin a "doomsday chemical" or a "potential risk"? Is embryo research a means to "enhance" fertility or a way to "manipulate" persons? Is genetic engineering a "boon" to agriculture or "tampering" with genes? Is Prozac a "therapeutic medication" or a "mind-altering drug"? Are incidents of scientific fraud "inevitable" or "aberrant"? Selective use of adjectives can trivialize an event or render it important; marginalize some groups, empower others; define an issue as a problem or reduce it to a routine.

Nor are words and metaphors the only way in which journalists convey values. By selecting their stories out of an endless stream of events and issues, they define certain subjects and not others as newsworthy. By their choice of headlines and leads, they legitimize or criticize public policies. By their selection of details they equip readers to think about science and technology in specific ways.

I approach the study of science journalism from the assumption that public communication is shaped by cooperation and conflict among several communities, each operating in terms of its own needs, motivations, and constraints. Journalists, their editors, and scientists themselves all influence the presentation of science in the press. The images of science and technology conveyed to the public reflect the characteristics of the journalistic profession, the judgments of editors about what the public will buy, and, above all, the controls exercised by the scientific community.

Scientists communicate to one another through specialized journals, but they must rely on the media if they want to reach a wider public. Conversely, the press relies on scientists as a source of information about complex but newsworthy aspects of health, energy, and environmental affairs. This mutual reliance plays a particularly critical role in shaping science news. Thus in the chapters that follow I will suggest how science writing reflects the characteristics of both science and journalism as these two professions seek to control the agenda of public communication.

Chapters 2 through 4 explore the interests and assumptions shaping the recurring images used by journalists to describe the work of scientists, the effects of technology, and the problems of risk. In Chapter 5, I turn to the question of how media messages are received and what impact they are likely to have on the reader and on policy choices. Chapters 6 and 7 address the characteristics of journalism that help to perpetuate certain overarching themes and images, and the professional constraints, cultural biases, and editorial pressures that shape the selection of science news. These characteristics of journalism converge with the complexity of many areas of science and technology to reinforce the tendency of journalists to look to scientists as neutral sources of authority. Thus, I explore in Chapters 8 and 9 the influence of scientists themselves on science journalism as they seek to control the language and content of science and technology in the press. Finally, Chapter 10 explains some fundamental differences between the interdependent communities of science and journalism to suggest both the limits and the possibilities of improving science communication.

Control over the information and images, the values and views, the signs and symbols conveyed to the public is obviously an extremely sensitive issue in today's society. Industries, political institutions, professional associations, special interest groups, and aspiring individuals all want to manage the messages that enter the cultural arena through education, entertainment, and above all, the media. Scientists are no exception. Analyzing their relationship with the press is thus a means of shedding light on problems that are of concern to scientists in their quest for improved public understanding of science and greater public support, to journalists in their difficult and important profession, and to those of us who want to be fairly and accurately informed about complex technical matters that affect our lives.

2

THE SCIENTIFIC MYSTIQUE

The public conceives of science as a "variant of the black art," wrote the editor of the *Nation* in January 1902. To the public, scientists are wizards and magicians, socially isolated from the society: "The scientist appears akin to the medicine man . . . the multitude thinks of him as a being of quasisupernatural and romantic powers. . . . There is in all this little resemblance to Huxley's definition of science as simply organized and trained common sense."[1]

In the 1990s science still appears in the press as an arcane and incomprehensible subject, far from organized common sense. And scientists still appear to be remote but superior wizards, culturally isolated from the mainstream of society. Such heroic images are most apparent in press reports about prestigious scientists, especially Nobel laureates. But the mystique of science as a superior culture is also conveyed in the coverage of scientific theories, and even in stories about scientific misconduct and fraud. This distanced and lofty image is useful for a community seeking public funds with limited public accountability. But far from enhancing public understanding, such media images create a distance between scientists and the public that, paradoxically, obscures the importance of science and its critical effect on our daily lives.

The Scientist as Star

Each year the media devotes considerable attention to winners of the Nobel Prize. With stunning regularity, stories of the Nobelists focus on their national affiliations and stellar qualities, using language recycled from reports of the Olympic Games: "Another strong U.S. show"; "Americans again this year receive a healthy share of the Nobel Prizes"; "U.S. showed it is doing something right by scoring a near sweep of the Prize"; "The winning American style."

Just as the papers count Olympic medals, so they keep a running count of the Nobel awards: one year "Americans won eight of eleven"; in another, "Eight Americans were recognized, tying a record set in 1972." "Since 1941," *U.S. News and World Report* announced to its readers in 1980, "the U.S. has had 26 Nobel winners in science, more than double the number won by second place Britain." In 1993 Nobel coverage, a *Boston Globe* reporter worried that the "American dynasty may falter in the future because the prizes . . . were for work done in the 1970s when basic science had more government support."[2] These stories describe nations as rivals somehow competing in a Nobel race for national pride, ignoring the international cooperation that is supposed to, and often does, characterize scientific endeavors.

Following the style of sports writing or reports of the Academy Awards, journalists emphasize the honor, the glory, and the supreme achievement of the prize: "The most prestigious prizes in the world. . . . They bring instant fame, flooding winners with speaking invitations, job offers, book contracts, and honorary degrees," runs a typical comment. In 1979 *Time* printed a large picture of the gold medal: "The Nobel Prize Winners are called to Mount Olympus; the recipients have worldwide respect."

Local or regional papers cover Nobel winners like sports or movie stars, seeking to find a local angle, however remote. Consider Roald Hoffmann, who received the 1981 prize in chemistry. A Rochester, New York, paper mentioned that he was a "Kodak consultant." The *New York Daily News* reported that he had graduated from Stuyvesant High School in New York City and printed his picture from the high school yearbook. Columbia University claimed Hoffmann as one of 39 laureates among its former faculty and alumni.

In one important respect, though, reports of Nobel awards differ markedly from sports writing. Coverage of sports stars often includes analyses of their training, their techniques, and the details of their accomplishments. However, except in the *New York Times* and specialized science journals, stories about Nobel scientists seldom provide details on the nature of the prizewinner's research or its scientific significance. Rather, when the research is described, it appears as an esoteric, mysterious activity that is beyond the comprehension of normal human beings. "How many people could identify with Mr. Hoffman's lecture subject: 'coupling carbenes and carbynes on mono-, di-, and tri-nuclear transmission metal centers'"?[3] Reinforcing the presentation of science as abstruse are photographs of scientists standing before blackboards that are covered with complicated equations.

In interviews with journalists, scientists themselves reinforce the mystification of science by emphasizing the extraordinary complexity of their work. Physicist Val Fitch described his research: "It's really quite arcane." "I find it difficult to convey to my family just what it is I've been doing," said physicist James Cronin in an interview with a *Time* reporter in 1980. *Time* cited the words of a member of the Nobel committee to describe Cronin's work: "Only an Einstein could say what it means."[4]

Just as science is described as divorced from normal activity, so the scientist, at least the male scientist, is portrayed in popular newspapers and magazines as socially removed, above most normal human preoccupations—"on the mountaintop," as the 1990 winner in medicine was described. Science appears to be the activity of lonely geniuses whose success reflects their combination of inspiration and dedication to their work. One scientist sees "in a most passive looking object" a "veritable cauldron of activity" that the rest of us are unaware of. Another "stumbled" on his find but then spent a year "probing" for errors. A frequent image is the scientist who spends twelve hours a day, seven days a week, at his work. Prizewinning scientists are part of "an inner circle of scientific giants" who talk about science "the way other people talk about ball games." Being with them is "like sitting in a conversation with the angels." How will the prize affect the lives of such dedicated people? A journalist from *Time* notes that Sir Godfrey Hounsfield, one of the 1979 winners in medicine, "plans to put a laboratory in his living room."

Instances of prestigious scientists behaving like ordinary mortals are noted with an air of surprise. The caption for a *New York Times* photograph of Walter Gilbert (the Nobel Prize-winning chemist who gave up his chair at Harvard to run the firm Biogen) notes his managerial skills: "[These] should not be underestimated just because he has a Nobel Prize."[5] Writing of Nobel physicist Hans Bethe's concern about the buildup of nuclear weapons, a *New York Times* reporter remarks that "he ultimately places his faith not in technology but in human beings—a remarkable stance for a man who has dedicated his life to the pursuit of science."[6]

While successful male scientists appear in the press as above the mundane world and totally absorbed in their work, the very few women laureates have enjoyed a very different image. Stories of female Nobel Prize winners appear not only in the science pages, but in such life style and women's magazines as *Vogue* and *McCall's,* which seldom cover science news.

McCall's described Maria Mayer, who shared the physics prize in 1963 for her theoretical work on the structure of the nucleus, as a "tiny, shy, touchingly devoted wife and mother," a woman "who makes people very happy at her home." Approaching the story in a personal style hard to imagine in the coverage of a male Nobelist, the reporter found Mayer "almost too good to be true"—"a brilliant scientist, her children were perfectly darling, and she was so darned pretty that it all seemed unfair." The reporter noted her "graceful union between science and femininity," but also emphasized the conflict between being a mother and a scientist, the guilt, the opportunities missed by not spending more time at home. Writing about Mayer's research, the journalist remarked that she explains it in a "startlingly feminine way" because she used the image of onion layers to describe the structure of the atom. The article then goes on to describe Mayer's reputation as "the faculty's most elegant hostess."[7]

Specialized science writers used similar stereotypes in their descriptions. A *Science Digest* article called "At Home with Maria Mayer" begins: "The first [American] woman to win a Nobel Prize in science is a scientist and a wife." It showed a picture of her, not at the blackboard, but at her kitchen stove.[8] Similarly, the *New York Times* headlined its article on Dorothy Hodgkin, who shared the prize for chemistry in 1966: "British Grandmother Wins the Prize."

The feminist movement did little to dispel such stereotypes. In 1977 Rosalyn Yalow, winner of the prize in medicine, also received extensive coverage in women's magazines. By *Vogue* she was characterized as "a wonderwoman, remarkable, able to do everything, who works 70 hours a week, who keeps a kosher kitchen, who is a happily married, rather conventional wife and mother."[9] *Family Health,* a magazine that reaches 5.3 million readers, headlined an article: "She Cooks, She Cleans, She Wins the Nobel Prize" and introduced Yalow as a "Bronx housewife." This journalist expected to meet "a crisp, efficient, no nonsense type" but discovered that Dr. Yalow "looked as though she would be at home selling brownies for the PTA fund raiser."[10]

Journalists had more difficulty fitting Barbara McClintock, recipient of the 1983 Nobel Prize in medicine, into this stereotype, and this in itself became news. *Newsweek* called her "the Greta Garbo of genetics. At 81 she has never married, always preferring to be alone."[11] (This article was published in the newsmagazine's "Transition" section—a column of briefly noted milestones, including weddings and deaths, among the noteworthy—although an item on McClintock also appeared on the "Medicine" page.) The *New York Times* covered McClintock in a long feature article. Its very first paragraph observed that she is well known for baking with black walnuts.[12]

The overwhelming message in these popular press accounts is that the successful woman scientist must have the ability to do everything—to be feminine, motherly, and to achieve as well. Far from being insulated and apart from ordinary mortals, women scientists are admired for fitting in and for balancing domestic with professional activities. As an exception to the usual coverage of scientists in the press, the portraits of female Nobelists only highlight the prevailing image of science as a recondite and superior profession, pointing up their lack of attention to scientific substance.

To complete the image of the esoteric scientist, journalists often convey the need of money and freedom to sustain science stars. During the 1970s, *Time* attributed the prominence of American Nobel Prize winners to the "heady air of freedom in U.S. academia and the abundant flow of grants. . . . Just do your own thing, the bounteous government seemed to say."[13] According to this 1979 analysis, the American tradition of freedom is far better for science than the "rigid" British, the "ideological" Soviets, and the "herr doctor" syndrome in Germany and France. Accepting the conventional (but questionable) wisdom among many scientists, this reporter expressed concern that the pressure to apply science to practical ends and to impose "cumbersome" regulations on experimental procedures will limit future triumphs. He also argued that U.S. science gains by insulation from humanistic pursuits: "The best minds have not been overburdened with required studies that are remote from their interests." Scientists, he emphasized, do such specialized and important work that they operate outside a common intellectual or cultural tradition.

The message of the 1970s was that science is distinct from cultural influence and that scientists need not consider the application of their work. In some respects this message has changed. Concerns about money and oppressive regulation persist, but in the 1990s media, the model scientist looks for practical applications and watches for ethical implications. In 1993, an admiring *New York Times* article on Francis Collins, director of the Human Genome Project, praised not only his vision and skills but his "personal empathy," his broad range of interests, and his concern about the applications and implications of genomics research. A quote from Sophocles shapes this report: "It is but sorrow to be wise when wisdom profits not."[14] However, the article also emphasizes that Collins is unique, a lonely genius who had been educated at home, able to work 100 hours a week with total dedication to "unlocking the secrets of the genome."

Ironically, even when treating scientists as removed from the common culture, journalists often turn them into authorities in areas well outside their professional competence. Thus we frequently read of their opinions on political dilemmas that have little to do with science. The implication is that science is a superior form of knowledge, so those who have reached its pinnacle must have some special insight into every problem.

The Authority of Scientific Theory

Although scientific findings are reported in the press, theories are seldom newsworthy. Notably excepted are those theories of behavior that bear on social stereotypes. Thus theories of evolutionary biology and natural selection, when used to explain human differences, have had a very active press. The theory of biological determinism attracted considerable news coverage following the controversy over psychologist Arthur Jensen's claims about the relationship between race and IQ. Its reappearance as sociobiology again attracted the press. The reports on sociobiology have been less concerned with substance than with purported applications. In selecting this subject for extensive coverage, journalists are in effect using a controversial theory to legitimize a particular point of view about the importance of biological determinism.

Sociobiology is a field devoted to the systematic study of the biological basis of social behavior. Its premise is that behavior is shaped primarily by genetic factors selected in a species over thousands of years for their survival value. Its most vocal proponent, E.O. Wilson of Harvard University, contends that genes create predispositions for certain types of behavior and that a full understanding of these genetic constraints is essential to intelligent social policy. He believes that sociobiology is "a new synthesis," offering a unified theory of human behavior. "The genes hold culture on a leash," he wrote in his book *On Human Nature*. "The leash is very long but inevitably values will be constrained in accordance with their effects on the human gene pool."[15]

Wilson's arguments about human behavior, extrapolated from his research on insect behavior, were widely attacked by other scientists for their apparent justification of racism and sexism, their lack of scientific support, and their simplistic presentation of the complex interaction of biological and social influences on human behavior.[16] Wilson's first book on the subject, *Sociobiology, A New*

Synthesis, was welcomed in 1980 by the *New York Times* as a "long-awaited definitive book." Subsequently, sociobiological concepts appeared in popular-press articles about the most diverse aspects of human behavior—used, for example, to explain:

- Child abuse: "The love of a parent has its roots in the fact that the child will reproduce the parent's genes." (*Family Week*)
- Machismo: "Machismo is biologically based and says in effect: 'I have good genes, let me mate'." (*Time*)
- Intelligence: "On the towel rack that we call our anatomy, nature appears to have hung his-and-hers brains." (*Boston Globe*)
- Promiscuity: "If you get caught fooling around, don't say the devil made you do it. It's your DNA." (*Playboy*)
- Selfishness: "Built into our genes to insure their individual reproduction." (*Psychology Today*)
- Rape: "Genetically programmed into male behavior." (*Science Digest*)
- Aggression: "Men are more genetically aggressive because they are more indispensable." (*Newsweek*)

The press is especially attracted to sociobiology's controversial implications for understanding stereotyped sex differences. The theory, we are told, directly challenges women's demands for equal rights, since differences between the sexes are innate. *Time,* for example, tells its readers that "Male displays and bravado, from antlers in deer and feather-ruffling in birds, to chest thumping in apes and humans, evolved as a reproductive strategy to impress females."[17] And a *Cosmopolitan* reporter, citing the "weight of scientific opinion" to legitimize his bias, writes: "Recent research has established beyond a doubt that males and females are born with a different set of instructions built into their genetic code."[18]

Sociobiological ideas easily shift to genetic explanations. The media have readily picked up on research suggesting a genetic basis of sex differences. In 1980, for example, two psychologists, Camille Benbow and Julian Stanley, published a research paper in *Science* on the differences in mathematical reasoning between boys and girls. Their study, examining the relation between Scholastic Aptitude Test scores and classroom work, found that differences in the classroom preparation of boys and girls were not responsible for differences in their test performance. The *Science* article qualified the implication of male superiority in mathematics: "It is probably an expression of a combination of both endogenous and exogenous variables. We recognize, however, that our data are consistent with numerous alternative hypotheses."[19] But the popular press was less circumspect, writing up the research as a strong confirmation of biological differences and a definitive challenge to the idea that differences in mathematical test scores are caused by social and cultural factors. The news peg was not the research, but its implications.

The original authors themselves encouraged this perspective, however, in their interviews with reporters, where they were less cautious than in their scientific writing. Indeed, they used the press to push their ideas as a useful basis on which to adjust public policy. According to the *New York Times,* they "urged educators to accept the possibility that something more than social factors may be responsible. . . . You can't brush the differences under the rug and ignore them."[20] The press was receptive. *Discover* reported that male superiority is so pronounced that "to some extent, it must be inborn."[21] *Time,* writing in 1980 about the "gender factor in math," summarized the findings: "Males might be naturally abler than females."[22] In 1992, *Time* continued to explain sex differences: "Nature is more important than nurture" and it is just a matter of time until scientists will prove it.[23]

In the 1990s, the ideas of sociobiology and genetics have been integrated into the media coverage of an extraordinary range of behaviors, including shyness, directional ability, aggressive personality, exhibitionism, homosexuality, dyslexia, job success, arson, political leanings, religiosity, infidelity, criminality, intelligence, social potency, and zest for life. These and other complex characteristics are treated as if they were simple Mendelian disorders, directly inherited like brown hair or blue eyes.

What is striking about the many articles on genetics and sociobiology is how easily reporters slide from noting a provocative theory to citing it as fact, even when they know that the supporting evidence may be flimsy. A remarkable article called "A Genetic Defense of the Free Market" that appeared in *Business Week* clearly illustrates this slide. While conceding that "there is no hard evidence to support the theory," the author wrote: "For better or worse, self-interest is a driving force in the economy because it is engrained in each individual's genes. . . . Government programs that force individuals to be less competitive and less selfish than they are genetically programmed to be are preordained to fail." The application of sociobiology that this author calls "bioeconomics" is controversial, he admits; nevertheless, it is "a powerful defense of Adam Smith's laissez-faire views."[24]

Similarly, a 1993 *New York Times* article is called: "Want a room with a view? Idea may be in the genes."[25] The reporter described "Biophilia . . . a genetically based emotional need to affiliate with the rest of the nature world." And in 1994, media attention is turning—albeit with caution and caveats—to controversial claims about an association between violent behavior and biological predisposition.[26]

The journalists who write about sociobiology recognize, indeed rely on, the existence of controversy to enliven their stories. Yet most articles convey a point of view by giving space to sociobiology's advocates and marginalizing critics, who are described as "few in number but vociferous" or people who are "unwilling to accept the truth."[27] In 1976 *Newsweek* suggested that Wilson was a victim like Galileo: "The critics are trying to suppress his views because they contradict contemporary ortho-doxies."[28] In 1982 *Science Digest* compared the criticism of socio-biology to the attack of religious fundamentalism on the theory of evolution ("Like the theory of evolution, sociobiology is often attacked and misinterpreted"[29])—a comparison that places socio-biology's scientific critics, such as Stephen J. Gould and Richard Lewontin, of Harvard University, in the same league as William Jennings Bryan.

In the 1990s, journalists are questioning the potential abuses of genetic information but not the benefits of the research or the credibility of research claims. We read of the importance of genetic explanations but then are warned that the ability to identify genetic predisposition is far ahead of therapeutic possi-bilities. Stories extoll the benefits of genetic research, but then decry the risk of gathering genetic information. We are told that this is the dawn of a new genetic era, yet cautioned about an impending eugenic nightmare. An article on the genetics of vio-lent behavior, for example, worries that people are "lulled into believing that all problems can be solved by science," but then raises few questions about the "new biological revelations."[30] A story about the "vexing pursuit of the cancer gene" laments the delays in a discovery promising therapeutic interventions, but then welcomes the delays as a "blessing" in light of the problems such knowledge will bring.[31]

The media response to sociobiology and, more recently, to genetic explanations of behavior, reflects the tendency to idealize science as an ultimate authority. By its selection of what theories to champion, the press in effect uses the imprimatur of science to support a particular world view. It does so, however, with little attention to the substance of science, its slow accumulative process, and the limits of its theories as an adequate explanation of complex human behavior.

The Purity of Science

Stories of fraud and misconduct in science appear in the media with increasing frequency. One would expect them to undercut the mystique of the authority of science, but this is not necessarily the case. On the contrary, journalists often report deviant behavior in a manner that further idealizes science as a pure and dispassionate profession.

Journalistic interest in scientific fraud began in the late 1970s as part of the post-Watergate preoccupation with corruption in American institutions. It has burgeoned in the 1990s as hundreds of articles report on incidents of data tampering, abuses of research subjects, and other forms of scientific misconduct. Some news articles on fraud are similar in style to reports of political or business scandals, describing particular acts of fraud, the investigations that revealed the incidents, and the institutional responses. Others discuss the issue of fraud more analytically, focusing on the causes and extent of misconduct, suggesting that particular cases are symptomatic of deeper problems. These two approaches reflect different interpretations. The first suggests that fraud is simply the deviant behavior of aberrant individuals, the second that it is a larger phenomenon with underlying causes basic to the present organization of science. Yet both convey a mystique about science, idealizing it as a sacrosanct, if vulnerable, profession.

Incidents of scientific misconduct elicit moral outrage, especially when research institutions are reluctant to recognize fraud or unwilling to react appropriately. This was the response of investigative reporters from the *Boston Globe* in covering a case of fraud at Boston University's hospital in 1980.[32] Marc J. Straus, a research physician specializing in lung cancer, had falsified data on his research subjects in order to claim the success of his research on cancer therapy. The *Globe*'s headlines focused attention on both the scandalous aspects of his behavior and the institutional failure to take action despite the seriousness of the offense, which involved human subjects. "Cancer Research Falsified," "Boston Project Collapses," "Doctor under Fire Gets a New Grant"—these articles emphasized that corruption in science was an unusual event.

Other reports of fraud use individual cases to criticize current research practices. A *New York Times* reporter, in a 1980 article entitled "The Doctor's World: How Honest is Medical Research?" called attention to competitive practices in research.[33] The *Christian Science Monitor* in 1982 interpreted fraud as part of "a larger problem," in particular the "corrosive effect of pressure to publish."[34] As incidents of fraud increased, journalists variously attributed the problem to "the pressure cooker of research," inadequate supervision of younger colleagues, or the fact that most experiments are never replicated because "you don't get a grant for repeating someone else's work."[35]

Changes in media interpretations of fraud appear when we compare the 1980 coverage of the Straus case at Boston University with the 1994 coverage of the falsification of evidence in a breast cancer study of the relative benefits of full vs. partial mastectomy. As in 1980, the disclosure of fraud brought weeks of outraged media coverage, condemning not only the incident, but also the National Cancer Institute for its extended delay in informing the public. Judging that the incident did not significantly change the research results, the NCI had simply reported the case in the Federal Register, hardly a widely circulated journal; physicians as well as patients found out about the incident much later from the press. Covering this case in the context of other incidents of fraud, journalists questioned the ability of scientific investigators to control large scale projects and of government to regulate them. Thus media disclosure of the incident brought about a profound sense of betrayal and mistrust.

In 1993, a research group at the Acadia Institute conducted a survey trying to quantify the prevalence and significance of scientific fraud.[36] Its findings—that problems were far more pervasive than many scientists believed and that the publicized cases were just the tip of a misconduct iceberg—attracted wide media attention. According to Lawrence Altman, M.D., the medical reporter of the *New York Times,* "It shattered the myth that fraud in science is a rarity." Articles today increasingly raise fundamental questions concerning the validity of certain traditional assumptions about science. Can scientific honesty be assumed? Is the scientific method adequate? Does the peer review process offer enough protection against fraud? But many reporters still avoid these structural issues, preferring to topple a few high-profile scientists through dramatic exposés.

Among the most visible cases involved the well-known AIDS researcher Robert C. Gallo. In 1989, John Crewdson, a reporter for the *Chicago Tribune,* accused Gallo of wrongly taking credit for discovering HIV, the virus that causes AIDS. The accusation led to a four year investigation involving the U.S. Congress, the National Institutes of Health, and Parisian researchers working on the same virus. Gallo was eventually cleared, but media reports throughout the ordeal were consistently hostile, turning Gallo, the leading American researcher on AIDS, into a villain, a symbol of "the desanctifying of the temple of science in the eyes of much of the public."[37]

Another much-publicized case involved Nobelist David Baltimore, president of Rockefeller University, who had been involved in the publication of a suspect paper written by a scientist in his research group. "Attack" journalism focused on individuals is relatively rare in the coverage of science, but Baltimore became a target. The media, dwelling less on the facts of the case than on his personality, portrayed the assistant who called attention to the incident as "powerless victim" with "selfless" motivations.[38] Reporters apparently took delight in toppling a star, described as "fall[ing] from a place of personal power." From most media accounts, readers would not have realized that Baltimore himself was never accused of fraud. Nor would they understand the problems of multiple authorship so prevalent in the biological sciences. The extensive coverage of this case failed to communicate the nature of scientific misconduct, behavior that can range from improper assignment of credit, to plagiarism, to the outright faking of data.

Whether journalists define fraud as an individual aberration or a growing problem in the contemporary practice of science, they project a coherent image of scientific ideals. The metaphors typically used to describe fraudulent data are instructive. They "contaminate," "tarnish," "besmirch," "taint," or "sully" the reputation of individual scientists, their institutions, and science itself. Faked data must be "expunged" or "purged." Scientists who learn that one of their colleagues is involved are reported to be "shocked," "horrified," "stunned," or "reluctant to believe it." For scientific fraud is a "sin" as well as a scandal. The culprit has "fallen," "betrayed" the profession. "When a scientist succumbs to temptation and pays the price, it is always sad."

Fraudulent acts in most other fields evoke quite different and less moralistic metaphors. Consumer fraud is a "ripoff" or a "crime," hardly a sin. Political scandals are abuses of trust and reported, often cynically, as critiques of political institutions. Bribery scandals or savings and loan company scams are covered as the corruption to be expected in local politics. The press is admired for exposing such incidents. But Crewdson's role in exposing Gallo was not admired by his journalist colleagues, who worried about diverting the scientist from his valuable research mission.[39]

Even in reports of scientific fraud, scientists are often idealized as dispassionate, objective, and with values that must remain above those in other fields. A *Newsweek* article clearly stated this view: "A perception of widespread fakery undermines the trust in others' work that is the foundation of science. More than business or law or politics, science rests on the presumption of honesty in a quest for truth. If that presumption comes into question, a backlash against science may not be far away. And that could compromise what is still one of the more objective and honest sources of information in an ever more complicated world."[40]

In the news reports of the 1990s, that backlash is increasingly apparent, but it has followed less from fraud than from growing realization that science is hardly independent of political and economic institutions. The exposés of Cold War radiation experiments (see Chapter 7) and reports on the commercial ties of scientists and academic institutions are changing public and media perspectives. Media attention is focused on the increased importance of industry-university collaborations, the "profiteering" of academic scientists, and the "gold-rush atmosphere" in several biomedical fields. Scientist are becoming "tycoons," "gene merchants," or "molecular millionaires." "Biology Loses her Virginity" reads a headline, as if science was ever so pure.[41]

Whether covering prizes, professional misconduct, or controversial scientific theories, the media convey a sense of awe about science. The scientist is portrayed as a star engaged in a competitive international race for prizes or prestige. Sometimes the intensity of competition can lead to misconduct, and sometimes research choices may compromise ethics. But science remains idealized as an esoteric activity, a separate culture, a profession apart from and above other human endeavors. This is a convenient image, serving the interest of scientists seeking status and autonomy, while allowing journalists to present problematic incidents as significant "news." But by neglecting the substance of science, ignoring the process of research, and avoiding questions of scientific responsibility, the press ultimately contributes to the obfuscation of science and helps to perpetuate the distance between science and the citizen.

3

THE PRESS ON THE TECHNOLOGICAL FRONTIER

In 1982, *U.S. News and World Report* looked ahead to developments that could be expected in applied science and technology. The story anticipated breakthroughs in many areas; "startling" progress in space age communication, "radical changes" in medicine, and "revolutionary" developments in agriculture, all with "far-reaching" influence on the way Americans live.[1] By the 1990s media projections were far more cautious. Reviewing the technologies to watch in 1994, the *New York Times* anticipated breakthroughs on the "digital frontier" among other areas, but observed that "a breakthrough technology is a breakthrough only if people adopt it, using it to change the way they work and live."[2]

While media projections vary in response to economic and social trends, they promote each new technology as the cutting edge of history, the frontier that will transform our lives. The frontier, however, is constantly relocated. In the 1960s space had been the new frontier; the names of the space probes were *Pioneer*, *Voyager*, and

Explorer. But space flights (until the *Challenger* accident) soon became routine—often in the news as much because of a misfire as for technological sophistication. In the 1980s frontier images appeared in the news coverage of computer advances, high temperature superconducting materials, and organ transplant techniques. In the 1990s, the frontier has moved to biotechnology, genetics, and cyberspace.

Every frontier has its dangers. Press reports can be apocalyptic when a new technology threatens prevailing values or when extravagant claims are not met. But most reports on technology are promotional, conveying the message that each new development will provide the magic to solve economic problems or ameliorate social ills. This promotional bias is most apparent in the coverage of computers and biotechnology.

Advertising High Technology

In the early 1970s the phrase "high technology" began to appear as a synonym for computer technology in specialized newspaper articles about investments in the computer industry. Soon "high technology" became a national symbol of progress, "as much a part of the political lexicon as motherhood, apple pie and the flag."[3] Hundreds of articles on high technology during the 1980s celebrated new computer developments as the "dawn of a new era," "the wave of the future," or "the force for revolutionary change." The images implied unlimited progress: "Experts believe that only economics and imagination limit the scope of computer technology: the revolution is real. . . . Every prediction is probably conservative."[4] Computer-based technologies were expected to resolve medical problems and even provide social success; one computer article promoted "high tech ways to meet a man." Articles on stock prices, product sales, computer camps, career choices, college enrollment, and military strategies refered to computers as the new frontier. By the 1990s the "frontier" extended to cyberspace, the "electronic superhighway," and the promise of "virtual reality"—even "virtual sex."[5]

The people who work along the high technology frontier are, in parallel terms, portrayed as pioneers or missionaries, forging "a new territory." A Minnesota computer consortium is "a pioneer in the midst of the high-technology prairie . . . blazing the trail for the educated around the world."[6] Cyberspace technologists are "network pioneers," "armed" with computers. Sometimes they are "gurus" or "apostles" and their followers "converts." In the 1980s Boston's Route 128 was "East Mecca," California's Silicon Valley, "West Mecca." Their products were "manmade miracles" or "economic magic."[7] With growing unemployment in the 1990s, the images have completely changed, replaced by stories of "job crunch" and images of "slamming doors."[8]

The language in high technology articles, however, remains focused on competition, struggle, and war—an imagery long used to promote science and technology. In November 1957 a *Newsweek* cover story was called the "World War of Science— How We're Mobilizing to Win It." Science was "the front"; "supremacy" over the "growing army of Soviet scientists" was the goal.[9] By the 1980s, Japan was "the enemy" in the technological arena. A *Newsweek* article on "the technological battle with the Japanese" pictured a Samurai warrior leaping out of a computer screen.[10] When the Japanese recession of the 1990s eased the competition, the media welcomed the "end of the bubble era," interpreting the declining interest in science among Japanese students as favorable to American interests.

Relying on corporate sources of information about new products, the media have adopted a corporate rhetoric, promoting applications and accepting, unreflectively, the assumptions of an aggressive industry seeking an expanded market. Articles appearing in the *Christian Science Monitor* during 1982 described computers that would provide reliable security systems, build appliances to simulate sight, hearing, and even human thought, create art, track the course of acid rain, link prison inmates to a career, offer a way out of the recession, search for people lost in a desert, help officials cope with emergencies, analyze poems and prose, aid the poor, provide hands-on experience in analyzing chemicals, teach complex repair tasks, make complicated concepts interesting to kids, solve crime, aid in the running of a restaurant, turn oil into a renewable resource, and bolster local economies.

A decade later, sophisticated data banks, computer graphics, and techniques such as DNA fingerprinting turned many of these promised applications into realities. However, some are problematic, threatening privacy and personal freedom. The threat became a media issue when the Clinton administration proposed to place on every telephone and computer an encryption device that would be accessible to the FBI. This became known in the press as the "clipper chip" that would "clip" the wings of individual liberty by allowing government to spy on private lives.

Economic changes also brought a shift in the media view of high technology. During the 1970s, few journalists sought to warn their readers that high technology prospects might fail. By 1983, however, such failures entered the news, accompanied by references to "cruel illusions" or "naïve exaggerations." A New York Times reporter pointed out that high technology can contribute to unemployment, create "mind-stunting, mind-dulling" jobs, and even encourage new forms of crime.[11] Newsweek revised its earlier oversell to mock the "visions of Atari Democrats who seem to believe that high-tech companies will be the country's economic salvation."[12] (That February, Atari had laid off 1700 workers and shifted its manufacturing operation from Silicon Valley to Taiwan and Hong Kong.) In 1985, however, the development of supercomputers brought a new round of optimism and promotion. The press welcomed federally funded supercomputer centers as a symbol of the United States's technological muscle, recording the enthusiastic words of supercomputer advocates, who promised no less than a "second renaissance."[13] But again, by the 1990s, the press reported that the computer "boom" had shifted to developing countries: "A Silicon Valley grows 500 miles from Bombay."[14]

In the early struggle for technological ascendance, there was little critical analysis of the potential social and economic problems of a high-technology society, such as worker displacement or the limited number of potential jobs. The economic changes of the 1990s brought some expression of disillusionment. But by this time media attention had already shifted to another area of science—the biotechnology frontier.[15]

Biology, the Belle of the Ball

"After years of being a dowdy old lady, biology has become belle of the ball."[16] Its revolutionary potential has attracted researchers "in droves," and "bankers [are] in hot pursuit." To the media in the early 1980s biotechnology was expected to produce the next economic miracle.

Ironically, the first reports on biotechnology were more about risks than revolutions. In the mid-1970s molecular biologists held an international meeting at the Asilomar conference center in California to assess the potential risks of recombinant DNA research. This was mainly a technical discussion, but the press evoked images of Frankenstein monsters and Andromeda-like strains spreading incurable disease. Some reporters worried about "warping the genetic endowment of the human race"; others about "biological holocaust." The message? Runaway science needs to be controlled.[17]

Only a few years later, questions of safety ceased to be news. Journalists dropped the subject, turning their attention to applications. Techniques of gene splicing, once represented as dangerous, became "a mundane tool," and headlines began to tout potential applications of the research as "miracles."[18] In 1977 *Time* magazine had run a cover story called "The DNA Furor: Tinkering with Life." Only three years later its cover story was called "DNA:

New Miracle." In 1976 the *New York Times Magazine* published an article called "New Strains of Life or Death?" A 1980 article in the same section was called "Gene Splicing: The Race toward Better Human Health."[19]

In the 1990s the "runaway science of genetic engineering" has become another "technological frontier." We read about the patenting of genetically engineered products, the implications of these products for resolving medical, agricultural, and industrial problems, the proliferation of genetics research and development firms, and, in local papers, the importance of these firms to the regional economy."[20] Journalists describe geneticists as pioneers, "unlocking the basic laws of nature," discovering the "secrets of life," solving the problems of devastating disease. The biotechnologists working on "high tech veggies" will "do wonders" to help meet nutritional requirements and to enhance the economy. The images that had characterized the reporting in computer developments have reappeared. Genomics researchers are "riding the DNA trail." Geneticists are "relentless hunters" involved in a "race" to find the markers for disease. And, cynically, they are directly compared to "Silicon Valley pirates trying to steal computer secrets."[21]

The media enthusiasm about biotechnology, however, is often tempered by expressions of doubt. Reporters have been especially critical of the growing links between the biotechnology industry and universities for their effect on open research, calling attention to the conflicts of interest that are endemic to this field where "profits and ethics collide."[22] The media have amplified the antiscience views of activist Jeremy Rifkin and the animal rights campaigns against the creation of transgenic animals. And with the rapid development of genetic testing, media attention is turning to the issue of genetic discrimination as tests reveal information about individual predispositions that could influence access to insurance or jobs.[23]

Reporting on biotechnology, the press bats readers back and forth from biotechnology miracles to visions of apocalypse, from celebrations of progress to warnings of peril, from optimism to doubt. These dramatic shifts take place as journalists respond to the promotional enthusiasm of technical institutions and then to changing popular fashions. But this fickle style of reporting also serves the interests of the media in their endless search for drama, for exciting "news." The same polarized style has marked the coverage of technological "panaceas"—organ transplantation, psychosurgery, estrogen replacement therapy, and reproductive technologies.

Problems and Promises of the Technological Fix

Media coverage of transplant techniques began in the 1950s with reports of a "revolutionary new stage of medicine . . . nearly ready to emerge from the research laboratory." The revolution arrived in 1968 when Dr. Christiaan Barnard's first heart transplant in Capetown, South Africa, stirred the public's imagination and attracted a flood of favorable and flamboyant reports. Journalists hailed Dr. Barnard's operation as a "surgical landmark," a pioneering venture comparable to space exploration. Barnard's work was described as a "revolutionary development bound to change our lives." Barely mentioned was the fact that his patient died.[24] Subsequently, the press heralded other heart transplant operations with front-page headlines but few technical details. Primed by press releases from the participating medical centers, reporters conveyed the misleading perception that the procedure was a miraculously effective solution for heart patients; they gave little attention to the patients' postoperative histories. The transplant was a dramatic event; the aftermath ceased to be news.

By the early 1970s, however, the problems of organ transplantation had become as newsworthy as the progress. Headlines announced "Heart Transplant Future Looks Bleak" and "An Era of Medical Failure." The media kept a box score of successful and failed transplant operations, and they publicized debates between the two Texas heart surgeons Denton Cooley and Michael DeBakey, who were respectively billed as "Texas Tornado and Dr. Wonderful." Reporters wrote of "transplant furor" and the new "medical rage."[25]

Then, in 1982, optimistic images reappeared in response to promotion of the artificial heart. The first operation implanting the Jarvik-7 model in a human subject, Dr. Barney Clark, was an experiment in a fishbowl. Reporters came to the Utah Medical Center and remained for the entire 112 days until Clark's death. The medical team included public relations experts who provided journalists with technical details and information on medical complications. However, according to *New York Times* reporter Lawrence Altman, much important information that would have better educated the public remained undisclosed.[26] Altman wanted to know how the patient was selected and the Jarvik-7 model chosen, and how the university's Institutional Review Board, set up to handle ethical dilemmas, had entered the deliberations. And he wanted to learn what scientific information was actually gained from the experiment. He felt that public relations control over the release of information limited the media's access to sensitive issues.

Other reporters, however, optimistically welcomed the experiment as a "dazzling technical achievement," "an astounding medical advance," "the blazing of a new path," a "medical milestone."[27] They turned the patient into a hero: "This man is no different than Columbus or the pioneers who settled this valley. He is striking out into new territory."[28] Frontier images mixed with military metaphors. Technology became a "weapon in the conquest of heart disease. New drugs that facilitated the operation were "weapons in the counterattack" against the resistance of the immune system. The medical technologists were "sleuths of the cardiovascular world," "men who race with death."[29] By focusing on the heroics of the procedure and avoiding substantive questions that would inform readers about the choices involved, the media provided readers with more drama than perspective. Their promotional view reflected reliance on the public relations advocates supplied by the hospitals; reporters did little more than regurgitate their claims.

The risk of relying on public relations sources became apparent in the press coverage of a different technology, estrogen replacement therapy (ERT). The story of these reports shows how promotional efforts to influence the press are most effective when they converge with journalists' perceptions of prevailing social values—in this case, the popular fantasy of remaining forever young.

ERT allays some of the symptoms of menopause and reduces osteoporosis, the thinning of bone mass characteristic of post-menopausal women. The therapy has been in the news since 1963 when the press proclaimed its extraordinary benefits and promised miracles.[30] Typical headlines read: "Preventing Menopause," "Science Paints Bright Picture for Older Women." An Associated Press (AP) newswriter cited as fact a scientist's assertion that "there is no reason why they [women] should grow old." Other newspapers and women's magazines termed the reluctance of some physicians to prescribe estrogen "archaic," claiming that "there is no scientific reason to object to its administration." The hormone treatment was touted uncritically as an exciting new discovery, a scientific fact, a cure for aging.[31]

Who were the experts cited by reporters promoting ERT? The major source of information was Dr. Robert A. Wilson, a gynecologist, an active promoter of estrogen treatment, and the director of the Wilson Research Foundation. Funded by three drug firms, his foundation existed to publish and distribute reports and recommendations about specific products. Wilson had published a popular book called *Feminine Forever*,[32] as well as a paper in the October 1963 issue of the *Journal of the American Medical Association* that described experiments using estrogen drugs to postpone menopause.

He and other advocates of estrogen treatment minimized the risks that cancer studies were making increasingly apparent. Mailing promotional materials on the rejuvenating effects of estrogen to newspapers and magazines throughout the country, they succeeded in attracting press attention. Articles promoting the therapy continued to appear despite a growing number of health warnings. An AP report in 1965, called "Pills for Femininity," cited a physician: "It can preserve the femininity of 17 million postmenopausal females in the United States. It would cost far less than a cocktail."[33] A 1966 AP article publicized the claims of an unspecified "medic" who promised eternal youth to takers of the drug; the article criticized doctors who worried about the side effects as "impediments to keeping women young and lovely.[34] And *Look* told its readers that "the vitality and freshness of the young girl need not fade at forty."[35]

Clearly the discovery of a pill that would keep women young was a newsworthy event. Reports of this discovery not only touched on a subject of wide interest, but conveyed a message readers wanted to hear. The problem with this promotional reporting was that it ignored or underplayed the growing evidence indicating ERT's potential risks. In 1969 the Food and Drug Administration (FDA) issued warnings about a relationship between estrogen and endometrial cancer.[36] And the next year, widely publicized U.S. Senate hearings on the safety of oral contraceptives called attention to the side effects of estrogen administration. The press duly reported that using estrogen for birth control could be harmful to health. But articles on using estrogen to postpone menopause selectively cited evidence to minimize the hazards of ERT and continued to emphasize its promise: "You can stop worrying about the menopause," said *McCall's* in 1971.[37] Still referring to Wilson's 1962 research, *Vogue* in 1974 promised "extra years of vitality," calling cancer worries "needless fear."[38] A 1975 article in the *New England Journal of Medicine* linking ERT to increased risk of endometrial cancer was likewise reported in the press, but so too were the persistent claims of ERT proponents. "When we drive the freeways we take a risk," a Beverly Hills gynecologist told Jane Brody of the *New York Times*.[39]

In December 1976 the public relations firm Hill and Knowlton proposed to Ayerst Laboratories, its client and a producer of estrogen drugs, a strategy for offsetting the warnings about the risk of endometrial cancer and for maintaining sales of estrogen replacement drugs. To "restore general perspective" and to "counteract unfavorable publicity," the firm recommended that Ayerst contact science editors. When doing so, the firm stressed, "it is important to steer clear of attempting to promote the use of estrogens, and instead concentrate on the menopause. . . . The estrogen message can be effectively conveyed by discrete references to products that your doctors may prescribe."[40]

Media interest in promoting a treatment to keep women "lovely and young" gave way to 1980s feminist sensibilities and the desire to be politically correct. But in the 1990s, as a baby boom generation reaches their fifties and public attention has turned to women's health, ERT is again in the news—despite the fact that there is little new to report on the costs and benefits of a medication sustained as much by ideology as by definitive research.

The influence of ideological fashion on media reporting is also apparent in the changes in the coverage of psychosurgical techniques during two periods, the 1940s and 1970s. In the early 1940s the press welcomed the new techniques of psychosurgery, then called lobotomy, with uncritical enthusiasm, though the medical community remained skeptical of the practice.[41] The press was responding to the promotional efforts of Walter Freeman, a neuropathologist who was advocating this dubious technique as a miracle cure for mental illness. Freeman had performed hundreds of lobotomies. They could produce changes in emotional responsiveness, and sometimes relief, but the costs in the impairment of the patients' intellectual abilities and judgment were often devastating. Although it could hardly be justified as a routine therapeutic procedure, Freeman solicited science writers to promote the technique. He staged psychosurgery exhibits for them at every American Medical Association meeting between 1936 and 1946. He invited journalists to his hospital to observe the surgery and provided them with success stories and testimonials from selected patients. Journalists paraphrased his words: Lobotomy was "no worse than removing a tooth"; "Surgeon's knife restores sanity to nerve victims." Publicity surrounded psychosurgical operations, with journalists welcoming lobotomy as "a promising cure for intractable problems."[42]

Titles of magazine articles listed in the *Reader's Guide to Periodical Literature* between 1945 and 1952 suggested that lobotomy was a means of cutting out cares, relieving unbearable pain, checking feeblemindedness, curbing psychosis, helping schizophrenics, solving crime, and curing epilepsy. During this period the *Reader's Guide* listed 40 articles on lobotomy: 18 of the titles indicated that its results were promising, 14 were neutral and descriptive, 4 posed questions ("Kill or Cure?"), and only 4 titles indicated that lobotomy could cause functional brain damage. Most *New York Times* articles during the same period conveyed similar optimism: "A Convict Made Normal"; "A Possible Preventive for Moral Degeneracy." Of the 18 articles in the *Times,* 11 emphasized the promises of the technique, 3 were descriptive, and only 4 cited those doctors who were warning their colleagues about the damaging side effects that must be weighed in extending the use of this surgery.

With advances in drug therapy in the 1950s, lobotomies became less prevalent. The practice reemerged in the 1970s, when the refinement of surgical techniques led to sufficient professional approval to justify its use on a small number of very sick individuals. At this time the press, however, reflecting public concerns about abuse and responding to extreme proposals that psychosurgery could be a "cure" for overcrowded prisons or even ghetto rebellions, paid attention to the social, ethical, and political implications of the procedure. The predominant metaphor in the press was "Clockwork Orange," referring to Stanley Kubrick's popular film, in which scientists used behavior modification techniques to transform a vicious criminal into an abused hero.

Of the 21 *New York Times* articles published on the subject in 1973 and 1974, 12 were negative, using such words as "mutilation" and stressing the problems of human rights. Seven were descriptive, focused on the need for regulation or the ethical question of voluntary consent. Two articles weighed the pros and cons of the technique; none repeated the promises of the earlier period. Similarly, of the 21 articles on psychosurgery in the *Reader's Guide,* 12 had such negative headlines as "A Robot" and "Pacification of the Brain."

Like the reporting on estrogen replacement therapy, the press coverage of psychosurgery reflected the convergence of the promotional efforts of advocates and the reporters' vision of prevailing social views. But what happens when views on a subject are clearly divided? In reporting about new reproductive technologies, the media have amplified the conflicts that have marked every new development in this rapidly changing field.

In vitro fertilization (IVF) has been represented in the media as both a humane technological solution to the problem of infertility and a dehumanizing intrusion of technology into the natural process of reproduction. Optimistic announcements of a medical breakthrough are matched by negative images of eugenics. It is "a crack in the door to a Huxleyan vision of a Brave New World." The birth of Louise Brown, the first "test-tube baby," in England in 1978 spawned a deluge of articles describing the event as a "sensational obstetric event," a "major medical achievement," a "miracle of modern medicine."[43] But virtually every article conjured up images of a baby hatchery. Louise Brown's first cry was "heard around the Brave New World," according to *Newsweek*. An article in *Time* referred to "Orwell's baby farm" and to "mass-produced kids." The *New York Times* wrote of the "brave new baby." Journalists covered the event as both a miracle and a sign of apocalypse. They cited optimistic physicians ("a very encouraging, happy new tool") and skeptics ("the potential for misadventure is unlimited"), enthusiastic sources ("a dramatic breakthrough, on a par with the discovery of anesthetics and penicillin"), and cynics ("It's a cookbook thing, something frogs do in a dirty stream"). They dwelled on the conflicting views of scientists, parents, and theologians.[44]

The birth of America's first IVF baby in December 1982 brought another burst of publicity, but with a slant clearly influenced by national pride. Welcoming the event as an "American achievement," journalists called it "a test-tube triumph," "a miracle of love and science," and they called the infant "America's baby." While still playing up controversy, they now attributed it to pressure from extremists, members of the Moral Majority and right-to-life groups. Focusing on conflicting views, few articles on IVF dealt with the limited success rate of the technique or the high cost of procedures that fail. There was little critical analysis of the legal, social, and ethical dilemmas that were to dominate media coverage of reproductive technologies in the 1990s.

Recent developments in reproductive technology have involved practices that many view as troublesome if not bizarre; the use of surrogate mothers, the long-term freezing of human embryos, embryo cloning, the efforts to harvest eggs from aborted female fetuses, and the impregnation of postmenopausal women. While some of these techniques, such as surrogate motherhood or embryo cloning, are not necessarily scientifically innovative, they are presented as cutting edge technologies or "landmark" procedures. But they are also treated as problems and threats: thus media interest has focused on their moral and social implications. Reports about the impregnation of postmenopausal women, for example, raised alarms—as if a deluge of 60-year-old women could be expected to become pregnant. Articles on the possibility of harvesting fetal eggs speculated on the reactions of a child who might discover one day that her "mother" was a dead fetus.

The most remarkable responses arose in coverage of a 1993 experiment carried on at a George Washington University laboratory, and intended to create additional embryos for IVF. The researchers had "twinned" a nonviable human embryo, but their experiment was covered—by major newspapers, magazines, and talks shows—as if it had yielded a cloning technology for the mass production of human beings. Envisioned were embryo and selective breeding factories, cloning on consumer demand, the breeding of children as organ donors, an industry of cloning and selling multiples of human beings, and even a "freezer section of the biomarket."[45] *Time,* for example, headlining its article: "Cloning: Where do We Draw the Line?" wrote of the "Brave New World of cooky cutter humans."[46] And the *Los Angeles Times* announced that "infertility, virginity and menopause are no longer bars to pregnancy."[47] Cloning, like every new technology, is expected to transform our lives.

The media coverage of new technologies shifts with prevailing fashions, but it plays on the desire for easy solutions to economic, social, or medical problems. Just as high technology is a "solution" to international competition, so medical technologies are "solutions" to problems of physical health, mental illness, or infertility. As we have seen, this style of reporting reflects the pressures from aggressive sources of information, as well as current fashions and editorial perceptions of what readers want to hear. Journalists respond to the interests of academic, industrial, and research institutions eager to promote the latest technologies and therapeutic techniques. But there is a counterside to these promotional tendencies, for exaggerated promises that dovetail with prevailing hopes or beliefs open the way to disillusionment, should they falter. The result is the tendency toward polarized reporting so evident in the coverage of technological risks.

4

THE PERILS
OF PROGRESS

Yesterday's technological "breakthroughs" often return to haunt the present. In the early 1960s rapid advances in science and technology brought speculations in the news about the "too-rapid pace of discovery" and the "mixed blessings of science." Ten years later the environmental and consumer movements focused these speculations on the potential risks to human health and survival posed by the products of technology. By the 1990s, warnings of the perils of progress have extended to global catastrophe, and attention has turned to ethical as well as physical risks.

The reporting of risk is a difficult area of science writing. Often selection of stories depends more on the potential for drama than the actual seriousness of risk.[1] Thus, news stories focus on catastrophic events. But in dealing with catastrophes, journalists must cope with complex and uncertain technical information, conflicting scientific interpretations, and social biases that affect perceptions of risk.[2] Norms of objectivity and fairness encourage reporters to balance different views—to give a technology's critics and proponents equal time—but such efforts expose them to criticisms from all sides. Industry groups and some scientists accuse reporters of taking a biased, sensational, antitechnology approach to reporting risks; they blame the press for creating unwarranted fear of technology—cancerphobia, chemophobia—and mistrust of industrial practices. Other scientists, environmentalists, and consumer advocates accuse the press of relying unfairly and almost exclusively on "establishment" expertise and of burying stories that might challenge local industries.

The coverage of controversies over the effect of fluorocarbons on ozone in the atmosphere, the ban on artificial sweeteners, the health effects of dioxin, and the risks of biotechnology illustrates several patterns in the reporting on risk. Attracted to catastrophic incidents, journalists emphasize competing interests, disputed data, and conflicting judgments, and then they turn to science as the source of authoritative evidence and definitive solutions—as the arbiter of truth. Trying to balance opposing positions, the media seldom explore the scientific issues involved in risk disputes or the methods of risk analysis that would provide a basis for meaningful judgments about competing claims.

The Ozone Controversy

The global dimensions of atmospheric pollution have been an important focus of environmental reporting since the ozone controversy began in the 1970s. This dispute attracted considerable media coverage, for at stake was atmospheric disaster resulting from the chlorofluorocarbons (CFCs) used as propellants in aerosol spray cans and as coolants in refrigeration. The release of CFCs into the atmosphere, scientists claimed, was depleting the layer of ozone that shields the earth from harmful solar radiation. Press reports on "the spray can war" exposed the public to a scientific dispute over an issue that few people at that time had ever heard about.

In 1973 British chemists published a paper in *Nature* warning that the release of fluorocarbons might damage the ozone layer.[3] A Swedish science journalist wrote a news article based on the paper, but newspapers in the United States failed to pick up on the story, labeling it as one more doomsday report.[4] Then, in 1974 two University of California chemists, Mario J. Molina and F. Sherwood Rowland, described similar findings at a meeting of the American Chemical Society.[5] The story took off when Walter Sullivan wrote an article for the front page of the *New York Times* explaining the dimensions of the problem. By publishing first and prominently, the *Times* in effect defined this risk as news, and articles followed in newspapers throughout the country.

The press coverage was apocalyptic and sensational, suggesting global catastrophe. The *Philadelphia Inquirer* wrote that "the world will end, not with a whimper but with a quiet p-s-s-t. . . . The earth may have already committed partial suicide or at least severe self-mutilation." The *New Haven Register* described an industry spokesman in the following terms: "Beneath the guise of Kris Kringle affability lurks one of the men who may cause the end of our present day earth." An AP dispatch asked, "Is a homely aerosol spray can and its charge of propellant gas sowing the seeds of doomsday?"[6]

Chemical industry executives responded by criticizing the press as a "dupe of environmentalists, consumerists and antiindustry scientists" and debunking the evidence from the Molina-Rowland study. The *Wall Street Journal* quoted an industry adviser as saying that the ozone depletion theory "resembles a structure held together by Tinker Toys, Scotch tape, and rubber bands."[7] *Business Week* quoted a chemist with approval: "If a ban were declared on fluorocarbons, we would have to go back to delivering blocks of ice to refrigerate food"[8]—a message borrowed from the nuclear industry ("if nuclear power is not developed, we will all freeze in the dark"). Du Pont responded by organizing a public relations program, inundating reporters with press kits and "clarifications."

In 1976 new data underscored uncertainties about the extent of ozone depletion. The ensuing controversy among scientists led to confusion as journalists struggled to understand the rapidly changing technical arguments. A report by the National Academy of Sciences (NAS) contributed to the confusion. The report confirmed the scientific studies warning that CFCs cause harmful deterioration of the ozone layer, though uncertainties remained about the severity of the effect. However, concerned about economic repercussions, the NAS cautioned against the imposition of regulation until there had been further research.[9] The ambiguities in this report left room for widely differing interpretations. The *New York Times* headlined its account of the report "Scientists Back New Aerosol Curbs to Protect Ozone in Atmosphere," while the *Washington Post* headlined its article "Aerosol Ban Opposed by Science Unit."[10]

Some inconsistencies in media coverage followed from efforts to give stories a local slant. In Orange County, California, home of fluorocarbon critic Sherwood Roland, a local paper ran a story on the NAS report under the headline "Aerosol Spewing Earth's Death," while a paper in Wilmington, Delaware, home of Du Pont, reported that "Ozone Study Gives Du Pont a Reprieve." But mostly the coverage of the ozone dispute reflected the indecision of scientists in response to rapidly changing evidence and their concerns about the economic impact of the recommendations.

Confused by complexity, most reporters remained silent on the nature of the evidence and the substance of the scientific dispute, simply balancing the views of contradictory sources. But the very existence of a publicized dispute had its effect. By late 1976, according to a survey conducted for the Consumer Product Safety Commission, 73.5 percent of the public had heard of the ozone issue, mainly via the press, and more than half had decreased their use of aerosol cans.

In 1978, the Environmental Protection Agency imposed a ban on the use of CFCs as aerosol propellants. For the press, this ended the newsworthy nature of the problem. Although the compound was still used extensively as a coolant in refrigeration and automobile air conditioners, as a foaming agent in the manufacture of polyurethane, and as a component in industrial solvents, the issue disappeared from the news. Thus, a 1980 EPA proposal for further controls received little media coverage, perhaps because the proposed regulation ran counter to the proindustry position of the Reagan presidential campaign. Industry was able to block regulation until the 1990s, when CFCs again became newsworthy following studies revealing higher than expected ozone depletion levels and warning of the effect on the incidence of human skin cancer.[11]

Media attention by this time had turned to the greenhouse effect and the global dimensions of atmospheric pollution. Reviewing the coverage of these issues, sociologist Lee Wilkins found three tendencies: an emphasis on the importance of progress, a promise that science would resolve the problems through technological solutions, and a focus on the opinions of experts. Journalists seldom cited environmentalists, manufacturers, or other interest groups on these global questions.[12] In some cases, however, the difficulties of dealing substantively with the scientific basis of a risk controversy has increased reliance on those sources most interested in biasing the news. This was the case in a series of disputes over the risk of consuming artificial sweeteners.

The Sweetener Dispute

"Bitterness about Sweets," "A Fatter Outlook for Diet Food," "Sweeteners Take Their Lumps"—when cyclamates in 1969 and saccharin in 1977 were found to cause cancer in laboratory animals, the press coverage was spiced with puns (often in poor taste). The 1958 Delaney Amendment to the Food, Drug, and Cosmetic Act required the Food and Drug Administration (FDA) to ban any food additives that were found in laboratory tests to induce cancer in animals or man. But the potential cost to the diet food industry turned proposals to ban sweeteners into public disputes. Two issues dominated these disputes: how to establish conclusive evidence about risk and whether to ban the food additives immediately or wait for definitive evidence of harm. The press focused attention on the role of science as a basis of regulatory decisions and the validity and significance of the tests used to evaluate risk.

In the coverage of the sweetener controversy, artificial sweeteners were under suspicion of causing cancer, and scientists were the detectives investigating the allegations or the judges who would discover the truth. In October 1969, *Business Week* announced that the "battle royal over the safety of artificial sweeteners is expected to be resolved next month."[13] Ten years later, with readers still waiting for definitive answers, *Newsweek* announced an eighteen-month "crash effort" by scientists to "resolve the dilemma of food additives like saccharin once and for all."[14]

This image of science perpetuates several convenient myths: that science can provide definitive answers about risk, that "facts" speak for themselves rather than being open to interpretation, and that decisions about socially acceptable risks are scientific rather than political judgments. Yet the frequent coverage of disputes among scientists also contradicted this image. Drawn to debates, journalists raised questions about scientific objectivity, identifying participating scientists with vested interests in regulatory decisions. *Business Week* criticized the "environmental mutageneticists" for using cyclamates as a "ready platform" for their interests. *Time* referred to "Hertz-rent-a-scientists" supported by industry.[15] On the one hand, science in the press is a necessary guide to policy; on the other, it is a biased institution influenced by vested interests. While both aspects contain elements of truth, juxtaposing attacks on scientific objectivity with images of science as a neutral arbiter of truth conveys confused and contradictory messages.

The coverage of a 1979 policy report on saccharin and food safety prepared by the NAS's Institute of Medicine illustrates the confusion. The report was intended to help the FDA assess the risks and weigh them against benefits for purposes of regulation. Using the word "risk" to describe both the frequency of harm (the chance of harm from the substance) and severity (the extent of possible harm), the NAS report placed saccharin in a "moderate to high risk" category of food hazards. It also emphasized the lack of evidence supporting the dietary benefit of saccharin that could be weighed against the risk. Nevertheless, the NAS panel stopped short of recommending a saccharin ban.

Like the NAS ozone study, this report was ambiguous both in its use of the word "risk" and its waffling on policy recommendations. Not surprisingly, its ambiguity became "red meat for the carnivores of the news media."[16] The *New York Times* described the debate among panel members as a "standoff," noting the judgmental and emotional factors that enter into scientific evaluation when it serves the needs of public policy, and emphasizing disagreements within the group.[17]

The *Times* also interpreted the panel's language in lay terms to mean that saccharin was a "moderate cancer-causing agent." The scientists on the panel objected; they understood their technical terminology to refer to both the frequency and severity of potential harm, not the likelihood that saccharin would cause cancer. The confusion over language was understandable, but it became a point of tension, exacerbated when the *Times* interpreted the academy recommendation as favorable to industry: "People who feel life would hardly be worthwhile without calorie-free sweeteners can probably relax."[18]

Readers of the *Washington Post* were no more enlightened.[19] On one day the *Post* published a column by the president of the

Calorie Control Council (the trade association of the sweetener industry) claiming that the academy had found "no evidence" of any association between saccharin use and cancer. A few days later it published a statement by the panel's chairman, who suggested that saccharin is a potential cancer-causing substance and should be phased out in three years. In writing about the academy report, neither scientists nor reporters effectively communicated the problems of regulation under an outdated law or the difficulties of balancing costs and benefits for purposes of regulation.[20] Meanwhile, the diet food industry engaged in aggressive efforts to influence the press. In a widely circulated brochure, the Calorie Control Council encouraged consumers to exert political pressure against "penalizing millions of diabetics and weight-conscious Americans." Citing phrases from the NAS report out of context, the council's brochure claimed that the academy "highlighted the benefits of saccharin," when in fact it had listed the claimed benefits only in order to show the limited evidence supporting them.

The Calorie Control Council also took advantage of the public's technical innocence about the validity and usefulness of the animal tests that were used to evaluate carcinogenicity. The carcinogenicity of saccharin had been discovered in 1977 when a researcher, using the accepted experimental procedure, fed his rats a dose equivalent to 50 times the maximum level considered safe for human consumption. The rats developed bladder cancer. Because it would be prohibitively expensive to use hundreds of animals in a long-term, low-dose test, it is standard practice to administer high doses in order to increase the likelihood of obtaining statistically valid results; lower dose effects are then extrapolated from the data.

The council debunked this procedure, and many reporters followed suit without discussing the nature of animal testing. Relying on public relations materials, they described the research as "ridiculous," "incredible," and "absurd." Popular magazines scoffed at a dosage they claimed to be equivalent to drinking

1250 cans of diet soda a day. *Newsweek* quoted without comment a spokesman from the William Wrigley company: "They give enough to sink a battleship and call it dangerous."[21]

While expecting answers from science, the press did a poor job of explaining research procedures and scientific concepts, including the validity of animal tests and the nature of scientific uncertainty in risk analysis. Assuming the industry position, reporters questioned the FDA's credibility, calling the agency "arbitrary, illorganized, arrogant, inbred and inept."[22] As a focus for press criticism, the FDA played much the same role in the sweetener dispute that the Environmental Protection Agency (EPA) would play during the debate over dioxin only a few years later. In both cases, the issue of agency competence reflected growing public concerns about appropriate regulation of science and technology.

The Dioxin Debate

A syndicated cartoon pictures Noah's ark stranded on a knoll. An animal on the ark asks his mate: "Why can't we leave, now that the flood is over?" The answer: "Dioxin." The longstanding media interest in the risks associated with the disposal of toxic chemicals began with the controversy over Love Canal, Hooker Chemical Company's abandoned waste disposal site in upstate New York. Hooker had dumped about 21,000 tons of chemical wastes in this canal between 1942 and 1953, then covered the dump and sold it for a token payment to the Niagara school board. A neighborhood developed on the site. The local press, in a town heavily dominated by the chemical industry, for years played down complaints, but in 1976 investigative reporters on the *Niagara Gazette* opened the issue to public scrutiny by publishing reports of heavy concentrations of chemicals in the area.[23]

The attention of the local press prompted a state investigation and a health alert that closed the school, evacuated some residents, and attracted national media attention. Love Canal became both a local disaster and a symbol for the growing problem of toxic waste disposal. The human interest angle of a child's death and a family's anguish, the David versus Goliath drama of local homeowners battling against state and federal bureaucracies, the anomaly of scientists scrapping in public about the severity of risks, the dramatic images of deformed babies and forbidden zones, and finally the realization of the national dimensions of the issue combined to make Love Canal a newsworthy event. By 1982 the press reported that 418 sites in the United States had been classified as dangerous toxic dumps.

This was the situation when dioxin was found in Times Beach, Missouri, in 1982. Again, complaints had been brought to the attention of journalists nearly a decade earlier, but the problem gained neither press nor official attention. Then, when a UPI correspondent published an internal EPA memo indicating many other potential toxic waste disposal sites, dioxin became news—a problem that could no longer be ignored.

The same pattern recurred in Woburn, Massachusetts, where in 1979 a local reporter had written a story about the high incidence of childhood leukemia near a toxic dump. There was no response; local fears were complicated by concerns about property values. When similar issues were on the national agenda in the late 1980s, the Woburn situation reached the national media with coverage on "NOVA," "60 Minutes," and in the *New York Times*.[24]

What images are conveyed in the reports on toxic wastes? At one extreme, sensationalized reports have portrayed dioxin as a major peril of technological progress. A 1980 *Time* cover story was called "The Poisoning of America." The cover showed a man swimming in a pool. Those parts of his face and body that were under the surface had dissolved to leave only the skeleton.[25] The *Los Angeles Times* in 1983 headlined an article "Lethal Dioxin Monster Guest in Chemical Lab." "Modern wizards of science with ultrasophisticated research tools," the reporter wrote, "have invented headache remedies, dandelion killers, and assorted bug poisons. They have also created a chemical monster." The reporter went on to compare Dow Chemical scientists to horror-movie chemists who accidentally create hideous mutants. Dioxin, he claimed, is "classified by scientists as the most deadly compound in nature after botulism and tetanus toxin."[26]

At the other extreme, the *Milwaukee Journal* described the dangers of dioxin as "overstated" and "unwarranted." "Is dioxin deadly?" asks the journalist. "Yes . . . for guinea pigs . . . and as far as is known now, not for humans." The article referred to "partisan politics and unsophisticated reporting" as causing "hysterical coverage." It quoted a "weary scientist" who did not want to be identified: "Dioxin is bad, dioxin is evil, dioxin is Darth Vader. Saying something good about dioxin is like saying Richard Nixon was a good father. It may be true but nobody would believe it."[27]

Reports on the health effects of toxic wastes dwell on the technical uncertainties that confound understanding of long-range risks. As in the coverage of artificial sweeteners, risk appears as a mystery to be resolved by a court of scientists. Dioxin is the suspect; science is the judge. But the scientists cited worked for the industries involved. Citing Dow Chemical scientists, the *Atlanta Journal* called an article, "Dioxin Studies Under Way but Verdict Is Not In."[28] Unsolved "mysteries" about the effect of dioxin were the theme of a 1983 *Arkansas Gazette* article citing Dow Chemical experts on tour as part of the firm's $3 million public relations campaign.[29]

Again reporters tried to deal with uncertainty about risk by balancing different points of view, but they provided little critical analysis to help readers weigh conflicting claims. A *Time* article, for example, quoted a Dow Chemical scientist ("no evidence"), the epidemiologist Irving Selikoff ("no question about it, dioxin is harmful to humans"), and a local housewife near a dioxin site ("almost everyone has thyroid problems"). Judgments of danger from those who were quoted ranged from "no risks" to

"Armageddon."[30] (Interestingly, of the hundreds of articles about the health fears of community residents near dioxin sites, few dealt with the workers in the chemical companies, who face similar risks every day.)

The 1992 election of President Bill Clinton with his environmentally concerned Vice President, Al Gore, renewed media interest in the dioxin issue, and optimistic news articles encouraged public expectations about better government regulation. But in less than a year, the tone had changed. The *Washington Post* described the EPA as evading responsibility in a case involving an incinerator in the Ohio River Valley.[31] The *New York Times,* describing a dispute over dredging dioxin from contaminated waterways, suggested that the government was "mired in muck."[32] Still, these problems are presented as aberrations, serious but limited questions of regulatory control. Few reporters seem inclined to raise fundamental questions about the nature of industrial practices or the environmental impacts of unfettered technological development.

Biotechnology Debates

Biotechnology applications have inspired futuristic risk reporting—speculations about the possible harm of bioengineered products yet to appear. One of the earliest disputes over biotechnology applications focused on the field testing of Ice Minus, genetically altered microbes intended to inhibit water crystallization and protect strawberries from frost injury. Environmental groups, concerned about health hazards, opposed these tests. Newspaper coverage highlighted the activities of Jeremy Rifkin, who has been a persistent critic of biotechnology since the recombinant DNA controversy in Cambridge, Massachusetts, in 1976. News reports on the Ice Minus field tests included striking and provocative photographs of the workers who were spraying the fields, wearing protective clothing that resembled the moon suits associated with the cleanup of toxic chemicals and nuclear wastes.[33]

Opposition to the bioengineered Flavr Savr tomato gained substantial media attention—as much, it seemed, on account of its irresistible potential for puns as for real evidence of risk. The genetically engineered tomato, introduced by the biotechnology firm Calgene in late 1991, was initially welcomed in the press as a fruit that would not rot on the way to the market. The product generated media stories on the "wonders" of high tech foods—leaner meat, celery sticks without strings, crisper and sweeter vegetables—and the press supported Calgene's effort to classify its product as a food rather than a drug that would be subject to FDA regulations. But then, as critics of biotechnology moved in, skepticism became fashionable, and journalists began to write about the tomato as a "frankenfood," a "killer tomato." There was a "tomato war" and a "tomatogate."[34] The idea of injecting mouse genes into food, the spectacle of chefs boycotting a tomato, the concern about "safe soup," attracted reporters who covered this product as an example of the risks that were bound to emerge from biotechnology. The business press responded by denouncing "crackpots and scaremongers" who hold back the "wheels of progress" by playing on public fears.[35]

The bioengineering of transgenic animals was especially controversial and a focus of media attention. Reporters addressed the concerns of small farmers who believed that biotechnology advances would give further economic advantage to agribusiness, and the opposition of religious groups who worried about the meaning of scientists tampering with nature and "playing God." The media uncritically followed the antics of animal rights groups who projected images of composite cattle, "geeps" (half goat, half sheep), and grossly oversized, distorted pigs. Here again, media-savvy Jeremy Rifkin was able to use the press to attract publicity for his anti-biotechnology campaign.

59

The health risks of biotechnology have never been established, and as yet there is little evidence they exist. Rather, the reporting in this area suggests that biotechnological risk is in many ways a surrogate issue, linked to deeper ethical and religious issues, concerns about economic inequities, and public mistrust. Indeed, the images pervading the media coverage of biotechnology are remarkably similar to those that had been projected during the nuclear power controversy—the synthetic monsters, the mutant animals, the mad scientists, and an industry out of control.

Risk reporting is widely criticized as hysterical, sensational, and confused. Clearly some reporting of risk is sensational, and many articles on this complex subject are confused and misinformed. But what are the sources of these problems?

The confusion characteristic of some (though by no means all) risk reporting in part reflects the journalistic response to fast-breaking events, technical obscurity, and conflicting information. When they report on catastrophic incidents, journalists have little time to do the research necessary to make an independent and probing analysis of rapidly unfolding events. For the most part, the media are reactive, relying on available sources of technical information. But these sources often perpetuate confusion. As we saw in the disputes over the health effects of dioxin, artificial sweeteners, and new biotechnology products, scientific disagreements, reflecting the limits of technical knowledge about health or environmental effects, contribute to confusion. So too do the efforts of interested groups and individuals, eager to project their own interpretations through the media. Applying naïve standards of objectivity, reporters deal with disagreement by simply "balancing" opposing views, an approach that does little to enhance public understanding of the role of science.

Though some risk issues require little technical background, few reporters try to explain to readers the nature of evidence necessary to evaluate human risk and the problems of judging how much evidence is necessary to warrant policy intervention. Perhaps more important, the press tends to reject statements by scientists who try to explain that they themselves do not know the extent of a given risk. In business to root out stories and expose potential coverups, reporters and their editors want definitive answers. They suspect that those scientists who claim lack of knowledge are trying to withhold information, to maintain secrecy, or to give them a "run-around." Seeking order and certainty, they convey the idea that science holds the solution to problems of risk, that "the acid test of scientific review" will resolve all uncertainties. But in doing so they perpetuate a false image of science, its contributions to the resolution of risk disputes and its limits as a basis for public policy decisions.

5

MEDIA MESSAGES, MEDIA EFFECTS

Journalists often cast the problems of technology in the form of a myth or social drama. Communities (Love Canal) are threatened by evil. Normal institutions (departments of health) fail to deal with the threat. Villains (polluting industries) are identified and brought into line through redressive action. Supervisory institutions (government agencies) are responsible for cleaning up. Solutions are sought in better technology (chemical waste containment facilities) or through scientific knowledge (expert advisory panels). The message is our ability to win over the forces that besiege us.

Prevailing values of order and efficiency shape the media coverage of many areas of science and technology. Scientific theories explaining human behavior are uncritically promoted in the press when they reaffirm comfortable social values. Scientific fraud is described as a violation, a moral affront to the purity of the social system of science. The risks from chemical waste disposal facilities or nuclear power stations are reported as problems of corporate or governmental negligence in a structurally sound system. Accidents, whether of super tankers or space ships, are blamed on aberrant individuals, sloppy corporate practices, or unfortunate decisions, and then euphemistically labeled "human error." Media stories seldom foster discussion of deep-seated structural dilemmas—fundamental problems in the organization and regulation of industry or distortions in the allocation of resources—that can lead to neglect of public safety. Interestingly, after the Chernobyl accident, the press did raise systemic questions about the Soviet system of managing the safety of nuclear plants, but such discussions seldom take place in the American context. Journalists, focusing on crises, are in "the outrage business."[1] Once a crisis is over, they revert to the language of cost-benefit analysis, avoiding broader questions about the value of efficiency or the human costs of technological change.

In the coverage of science, the media has encouraged the widely held belief that science is distinct from politics and beyond the clash of conflicting social values. Although individual scientists are sometimes criticized as biased, science as an institution is assumed to be a neutral source of authority, the engine of progress, the basis for just solutions in controversial public affairs. Seldom do science writers analyze the distribution of scientific resources, the social and political interests that control the use of science, or the limits of science as a basis for public decisions.

The images of science and technology in the press, as we have seen in Chapters 2 through 4, are often shifting, reflecting current fashions and prevailing fears. Today's exaggerated promises—of new fixes, new devices, new cures—become tomorrow's sensationalized problems. What is the effect of such images? How do readers respond? Does science journalism in fact shape, or even create, public attitudes—or does it simply mirror them? And what is the influence of media coverage on public policy?

63

Media Influence and Public Attitudes

Technological failures often seem to cluster, attracting an outraged press. The Three Mile Island accident, the falling of *Skylab,* the crash of a DC-10, and a series of automobile company recalls of defective parts all occurred in 1979. In July of that year a group of industry representatives, academic scientists, engineers, social scientists, and pollsters met to consider the impact that extensive press coverage of these events was having on public attitudes.[2] The participants worried that journalists were "smearing science" in their reporting of engineering accidents. They were, however, reassured by public attitude surveys indicating no significant change in behavior or loss of public confidence in science and technology, except in the case of specific technologies such as nuclear power. Rather, the surveys suggested remarkable continuity in the public's trust despite short-term fluctuations. Public attitudes toward science seemed to remain, in the words of one participant, "overwhelmingly favorable."

The 1980s saw another series of technological disasters—the Bhopal chemical spill, the explosion of the *Challenger* space shuttle, and the fire at the Soviet nuclear power reactor at Chernobyl. Again, the critical coverage of these accidents could have provoked a sweeping Luddite response, and surveys suggested a brief decline in public support of science; but after the initial shock following each event, public attitudes appeared to favor further development of the technologies, except for nuclear power.[3] News of disasters appears to have only limited effects on attitudes shaped by years of overwhelmingly favorable reports of the benefits of technology.

Social scientists have long assumed that media images matter, but views on their actual degree of influence have differed.[4] In the 1920s, communications researchers began to focus considerable attention on the media's power to shape public opinion and, in particular, to impose political ideas. Charles Cooley and other social scientists envisioned the press as an important source of community values, the basis of social consensus, the "thread that holds society together."[5] Walter Lippmann stated the prevailing belief: "The power to determine each day what shall seem important and what shall be neglected is power unlike any that has been exercised since the Pope lost his hold on the secular mind."[6]

In the late 1930s a growing number of studies challenged these assumptions, suggesting that the press was seldom the direct source of attitudinal or behavioral change. In light of this research, Paul Lazarsfeld and his colleagues developed a multistage model of communication effects in which information flows from the press to opinion leaders and through these leaders to the community. This model emphasized the mediating effect of personal relationships and primary groups such as the family in influencing people's response to media coverage.[7] During the mid 1950s, however, analysts began to argue that the United States had become a "mass society," characterized by atomization and the isolation of individuals. The declining importance of primary groups, they believed, enhanced the direct influence of the press.[8]

In his influential 1960 book on the effects of mass communication, Joseph Klapper summarized this research, arguing that the images conveyed by the press are assimilated and interpreted by different readers according to their prior beliefs, predispositions, personal experience, and the attitudes of their peers. In Klapper's view, press communication contributed to, but was not the primary cause of, the public's attitudes and ideas.[9] The effect of media messages, he argued, depends on the social context in which they are received. This perspective was to shape the assumptions underlying subsequent media research.

Many factors enter the reader's social context, including the cumulative influence of past media images and various alternative sources of information and imagery, such as fictional television stories, documentaries, comic strips, and other vehicles of popular culture.[10] In a content analysis of 1600 television programs broadcast between 1969 and 1979, George Gerbner, professor of communications at the University of Pennsylvania, found that science appeared less in television news than it did in entertainment and science-fiction programs.[11] These programs often focus on situations of crisis and danger, and they portray scientists as forbidding and strange. Similarly George Basalla, professor at the University of Delaware, has found that comic strips portray scientists as "remote, superbly dedicated, logical and humorless individuals, apt to put reason above human considerations." The stereotypical scientist works in isolation, beyond public scrutiny. He is "a Faust-like figure who has great power over natural forces. . . . Because of his evil or damaged nature or because of circumstances beyond his control, the power of science is likely to be used against mankind."[12] Such archetypal images are evoked by research in biotechnology and genetic engineering: the 1990s entertainment media is full of Frankensteins and Fausts. Popular films such as *Jurassic Park* or *The Fugitive,* portraying negative images of scientists as villains, are blockbuster hits. But popular culture is hardly homogeneous, and people are also exposed to medical melodramas portraying scientists as heroes and technological fantasies promoting the wonders and promises of science.

In a historical study of American popular magazines between 1915 and 1955, Marcel LaFollette of George Washington University found mainly inflated images: "Mass media discussion of science combined extensive reporting on the actual results of science, promises and predictions that science would cure any social problem, and images of scientists as omniscient, powerful, well-meaning, and heroic, to develop a climate of expectations of what science could do for and to society."[13]

Today, the images remain inflated, but they vary in different types of media. The ubiquitous "advisory arts" are a source of inflated expectations about what science can do. The extraordinarily popular "how-to" books and syndicated health columns provide endless information (sometimes misinformation) about health and science. Readers are presented with promises of future technological marvels and images of science as the ultimate source of authority. Similar promises are to be found in television documentaries, idealizing and explaining science with elegant visual images and starring authoritative scientists who assume the role of "host."

For many readers, their contact with "science" comes through articles in the tabloids so visible at the supermarket checkout counter. The *National Enquirer*, the *Weekly World News*, and the *Star* all devote a lot of space to scientific and medical events or pseudoevents—mostly as a form of entertainment and titillation. Their stories take the press's preoccupation with scientific wonders and breakthroughs to its logical, if extravagant, conclusion. Scientists in the tabloids "play God" in their "secret laboratories." They can "cure splitting headaches," "create human beings," "guarantee boy babies," "save brain-damaged victims," "cure impotence," "prevent cancer," and "find the key to long life."[14] Interestingly, in the age of genetic engineering and reproductive technology, such tabloid images are appearing as well in the mainstream news, as journalists speculate about the futuristic implications of contemporary research.

Tabloid images of science also appear in reports that border on pseudoscience and superstition. Many newspapers not only run a daily horoscope feature, but give prominent attention to astrological and pseudoscientific claims that purport to predict elections, natural disasters, economic trends, or personal affairs. Although such claims may be played down or dismissed, their frequent appearance in the press lends them an aura of credibility.[15] Indeed, portraying such claims as having "won" over science, John Burnham blames the media for antiscience thinking.[16]

About 100 million Americans follow the comic strips analyzed by George Basalla; millions read the tabloids, and most Americans watch many hours of television every day. Images appearing in these vehicles of popular culture help create the expectations individuals bring to their reading of science news. But their actual influence on attitudes and behavior remains disputed and is difficult to assess.

There is a great deal of public interest in science and technology, and scientific knowledge is considered both desirable and attainable. Surveys in 1988 by the National Science Board indicate that 84 percent of American adults disagreed with the idea that "it is not important for me to know about science in my daily life," and 72 percent agreed that "if scientific knowledge is explained clearly, most people will be able to understand it."[17] A 1993 Harris Poll found that four in ten American adults "have more than a passive involvement in seeking science news." Thirty-eight percent read science news in the newspapers at least once a week.[18] Sixty-nine percent of respondents believed that science news was as important as news about crime, and 63 percent felt it to be as important as politics. In particular, people welcome science news that relates to health and disease and to waste disposal, suggesting that science news is mainly valued in terms of its relationship to the problems of daily life. Many people, for example, use the media as their primary means of learning about ways to keep healthy and fit. A National Cancer Institute survey of how people become informed about ways to prevent cancer found that 63.6 percent get their information from magazines, 60 percent from newspapers, and 58.3 percent from television. Only 13 to 15 percent had talked to physicians about cancer prevention.[19]

Despite the interest expressed in media coverage of science, the actual influence of the media on beliefs and behavior varies with the selective interest and experience of readers. In areas of science and technology where readers have little direct information or preexisting knowledge to guide an independent evaluation (e.g., the effect of fluorocarbons on the atmosphere), the press, as the major source of information, defines the reality of the situation for them. But where readers already have an established set of biases (e.g., about the causes of social behavior), science reporting tends to justify and reinforce these biases. And when the reader has had personal experience (e.g., with workplace risks) or long-term exposure to media coverage (e.g., about environmental problems or dietary risks), the effect of media images is tempered by prior attitudes about the issues.[20]

Public beliefs about science and technology tend to correspond with the messages conveyed in the media, though the direction of cause and effect is not clear. Though the media are often blamed for biasing public opinion, journalists claim they are merely reflecting public views. In a study tracing the relationship between media discourse and public opinion about nuclear power, sociologists William Gamson and André Modigliani found an interactive pattern: "Media discourse is part of the process by which individuals construct meaning, and public opinion is part of the process by which journalists . . . develop and crystalize meaning."[21]

In part, the difficulty of assessing the influence of the media reflects the persistent contradictions in the public view of science. In 1965, historian Oscar Handlin suggested that science was hardly assimilated by the public even in the cultures of advanced industrial societies which expend enormous resources to support the scientific enterprise: "Paradoxically, the bubbling retort, the sparkling wires and the mysterious dials are often regarded as a source of grave threat. . . . The machine which was a product of science was also magic, understandable only in terms of what it did, not of how it worked. Hence, the lack of comprehension or of control; hence also the mixture of dread and anticipation."[22]

Conflicting attitudes persist. Gilbert Omenn, a science policy analyst at the University of Washington, found in 1983 that many youngsters perceive scientists as "geniuses or weirdos, people with whom they have little hope or no desire of identifying. . . . People close at hand are not recognized in their scientific roles. They don't fit the stereotype as brilliant problem solvers grappling heroically with the unknowns of nature."[23] Yet surveys also demonstrate a profound and unquestioning belief in the authority of science and the ability of scientists to make decisions in many areas of public policy.[24] Feelings of apathy, a sense of impotence, and a tendency to defer to expertise are all prevalent in the American population, as these surveys show. Their findings are certainly consonant with the media portrayal of science and technology as esoteric and arcane, yet also a source of the authority and expertise relevant to everyday problems.

Public acceptance of science appears to be based on expectations of immediate applications, promising solutions for growing energy costs, effective new therapies, efficient ways to handle toxic wastes. This emphasis on applications is encouraged by scientists themselves in their quest for public support. It is perpetuated in the popular media through their focus on short term implications. And it is manipulated by corporate advertisers ("It is a medical fact . . ."). But unrealistic public expectations about the benefits of science also leave the enterprise vulnerable when things somehow fail to "work."

Bad news about science and technology can influence consumer behavior, especially if alternative products are available. Thus, following the ozone controversy, people bought fewer aerosol sprays. And sales of Tylenol dramatically declined following a widely publicized tampering incident. After extensive media reports on dietary studies relating cholesterol-producing foods with heart disease, consumption of beef, eggs, and fatty milk products declined. News coverage of toxic shock syndrome adversely affected the sales of some brands of tampons. The media's exposés of leaky silicone breast implants led to demands for remedial surgery. And news reports about discoveries of a genetic predisposition to breast cancer brought women to clinics demanding genetic testing and even preventive mastectomies.

Media stories about risk can lead to dramatic changes in behavior even when data are speculative. But according to a 1993 study, public responses to new technologies have less to do with the balance of costs and benefits than the social issues associated with the risk; for example, the regulatory context, the ethical implications, and the general sense of how the technology is controlled.[25]

The influence of "good news" is apparent in the remarkable history of Prozac, an anti-depressant drug. In 1990, a spate of stories about Prozac appeared in the popular press. The "Promise of Prozac" was most prominent as a cover story in *Newsweek* that, instead of the usual celebrity, pictured a large green and white pill.[26] Overnight the drug became a star, appearing on talk shows, magazines, and news reports as the "feel good" drug. Taking Prozac meant "bye bye blues" according to the media, which reported little about what it did or how it worked. But there was soon a backlash, and the pill was blamed for suicides and violent behavior. When some patients or their families sued the manufacturer, Eli Lilly, Prozac became "a medicine that makes you kill." Thus, for several months the media described this medicine as both a "wonder drug" and a "killer drug."[27] The backlash, however, had little influence on growing consumer demand, and the media image improved. According to Peter Kramer, M.D., who traced the history of Prozac's fame in his best-selling book *Listening to Prozac,* the extraordinary media attention brought Prozac remarkable popularity. Prozac has become a legal drug of choice not only for depressives, but for normal people seeking a mind-altering drug, a technological fix[28]—so much so that in 1994 *Newsweek* again featured the green and white pill on its cover and ran a series of articles called "Beyond Prozac" about the future of "made to order personalities."[29] Quoting a neuropsychiatrist, the message was, "For the first time in human history we will be in a position to design our own brains." Tracing the rising popularity of Prozac, *Newsweek* said it now has "the familiarity of Kleenex and the social status of spring water." Sales, by the end of 1993, had grown to $1.2 billion a year.

The media coverage of Prozac had a major influence on its rapid rise as a drug of choice. However, deliberate efforts to use the press to influence behavior do not necessarily have the effect anticipated. Despite extensive news coverage of the Salk polio vaccine when it became available in the late 1950s, relatively few individuals agreed to be vaccinated at that time.[30] Similarly, media coverage of the 1964 U.S. Surgeon General's report on smoking and cancer had little direct effect on smoking habits in the short term.[31] More important were the long-term policy changes restricting smoking in public places, but these took over 20 years to be implemented. Although people seek information from the press to guide even the most personal decisions (such as choice of birth control technique), they use such information mainly when it corresponds to their prior views and inclinations.

A more general effect of press coverage is to establish a framework of expectations, so that isolated events take on meaning as public issues.[32] For example, before the nuclear power accident at Three Mile Island, a series of less critical accidents at other nuclear power plants appeared as isolated local events. The extensive press coverage of Three Mile Island alerted people to the larger context of such events and placed the issue of nuclear power plant safety on the public agenda. Similarly, before 1975 few Americans had even heard of dioxin or of a toxic waste disposal problem. By 1980, 64 percent of the public had heard of the issue and were worried about the disposal of industrial chemical wastes.[33] The press had made these problems visible and defined a "frame" within which they could be interpreted. In many such cases, the actual events have little immediate meaning for most individuals. It is the media that create the reality and set the public agenda, directly influencing policy decisions.

The Framing of Public Policy

By their selection of newsworthy events (e.g., a new AIDS therapy), journalists define pressing issues. By their focus on controversial problems (e.g., the location of toxic dumps), they stimulate demands for accountability, forcing policymakers to justify themselves to a larger public.[34] By their use of imagery, they help to create the judgmental biases that underlie public policy. The media can influence public policy even in areas where there is broad indifference on the part of the electorate. Indeed, the media's power to generate pressure for policy changes may be relatively independent of prevailing public attitudes. The real power of the press, claimed Harold Laski, comes from "its ability to surround facts by an environment of suggestion which, often half consciously, seeks its way into the minds of the reader and forms his premises for him."[35]

Metaphors in science journalism cluster and reinforce one another, creating powerful images that have strategic policy implications. When high technology is associated with "frontiers" to be maintained through "battles" or "struggles," the associations of war imply that the experts should not be questioned, that new technologies must go forward, and that limits are inappropriate. But if, instead, the imagery suggests internal peril, crisis, or technology out of control (as in the case of certain risks), then we seek ways to "rein in" the "runaway forces" through increased government regulation and control. Calling the weakness of science education a "problem of educational policy" implies the need for considered, long-term intervention; defining it as a "national crisis" implies the need for an urgent, if short-term, response. If science is incredibly complex and arcane and the scientist a kind of magician or priest, this implies that the appropriate public attitude is one of reverence and awe. But if science is simply another interest group seeking its share of public resources, the need implied is for critical public scrutiny.

73

The terms used to describe a problem, the sources cited, the perceptions conveyed, point the finger of blame and imply responsibility for remedial policies. To define the relative contribution of personal life style, genetic predisposition, or nearby chemical hazards to the unusual incidence of breast cancer among women in parts of Long Island influences judgments about liability and claims for compensation. To evaluate the relative significance of human error and equipment failure in causing an accident such as the *Exxon Valdez* supertanker oil spill has serious consequences in terms of who must pay. To place the blame for harm from exposure to toxic waste implies responsibility for cleaning up contaminated sites. And to attribute deviant behavior to individual genetic predisposition rather than to social or economic conditions limits the responsibility of public institutions.

Media reports have often directly influenced public policy. Extensive media coverage of disputes over waste disposal dumps helped to bring about changes in national policy and forced the reorganization and restaffing of the Environmental Protection Agency in the 1980s.[36] The media coverage of the controversy over recombinant DNA research led the mayor of Cambridge, Massachusetts, to organize a citizens review board to evaluate the wisdom of building a laboratory in the city.[37] Publicity surrounding the debate over laetrile as a cancer cure forced the National Cancer Institute to test the drug on human cancer patients even though previous failures to demonstrate therapeutic effects on animals would normally have precluded human testing.[38] The media coverage of demands from gay activists helped to convince the National Institutes of Health to release the AIDS drug AZT before the completion of clinical trials. And media outrage over the falsification of data in the research on alternative breast cancer therapies forced the National Cancer Institute, for the first time in its history, to take data out of the hands of a principle investigator for reanalysis.

By creating public issues out of events, the press can force regulatory agencies to action simply out of concern for their public image. Within four days of the initial press reports on experiments finding bladder malignancy in rats exposed to high doses of cyclamates, the commissioner of the Food and Drug Administration put in place the cyclamate ban—perhaps a record for the rapidity of an administrative response to a technical study. Why such haste on the basis of limited evidence? "We were afraid of a leak," said the commissioner. He feared that public exposure in the press would politicize the FDA's activities and bring about legislative oversight.[39] This, indeed, happened in the 1990s when NIH whistleblowers went to the media with stories of fraud in science: the disclosures led to Congressional investigations and greater public scrutiny of NIH procedures.

Media coverage can also influence the financial support given to research, a fact well understood by scientists and their institutions. In the 1940s the proliferation of cancer stories in the press helped convince Congress to give research support to the National Cancer Institute. The press dramatization of infantile paralysis in the 1950s attracted millions of dollars to the support of research in this area. Dramatic news stories on AIDS helped to generate public funds for AIDS research. Widely published reports of declining U.S. leadership in high technology in the 1980s influenced legislative decisions to support costly mega-science projects. Ten years later, media accounts of inefficiency contributed to the decline in federal support.

Corporate decision makers are also influenced by media coverage of science and technology, especially when issues lie outside their day-to-day sphere of competence. As their source of information, the press can lay the groundwork for establishing research directions and the credibility of requests for funding. The wave of interferon publicity in the late 1970s whetted the appetite of industry. During the peak of media hype, corporate and government spending on interferon studies greatly outpaced the overall rate of growth in biomedical research, even though the peer review ratings in this area were lower than in other fields.[40] The extensive media coverage of biotechnology suggested its growth potential, increasing the availability of venture capital for new biotechnology firms during the 1980s. Later, in the 1990s, media reports of public concerns about biotechnology encouraged caution. When the press reported the opposition to Calgene's Flavr Savr tomato, the price of the company's stock fell, and it temporarily took the product off the market. By amplifying possibilities and calling attention to potential problems, the media can influence the dissemination of new products and shape the direction of scientific and technological priorities.[41]

Publicity may not only affect the allocation of financial resources, it may also bring external controls and turn the interest of scientists toward or away from specific areas of research. The right-to-life protests against fetal research in the 1970s were widely reported in the press, and fear of harassment discouraged many scientists from working in this field, even before the government ban on such research in 1981. Interestingly, President Clinton's lifting of the fetal research ban in 1993 attracted far less media attention.[42]

The widespread reporting of the XYY controversy (studies examining the effect of having an extra Y chromosome on male criminal behavior) in the late 1970s also brought harassment of the scientists involved and discouraged further work in this area.[43] New studies of the genetics of violence in the 1990s again attracted the media. When the University of Maryland proposed a scholarly conference in 1993 to examine this research and received NIH funding, critics appeared on television, attacking the plan as supporting racist research. The result? The NIH withdrew its funds for the conference. Later, after an internal NIH committee concerned about academic freedom reviewed the issue, the agency reinstated the funding, and the conference was rescheduled for 1995 in the unlikely hope that the media would by then lose interest.

In light of their influence on public policy, the media today represent a battleground for political and economic interests seeking to convey their views to the public. As issues involving science and technology become increasingly at stake, scientists also must enter this arena. Their contacts with journalists are often strained, for the communities of science and journalism approach the problem of public communication from different professional perspectives, cultural frames, and political perspectives. Thus, to understand science journalism, we must look at the interaction of these two communities, each with its own normative assumptions, social biases, and professional constraints.

6

THE CULTURE OF SCIENCE JOURNALISM

A staff writer for a Joseph Pulitzer newspaper, the *Sunday World,* once described the philosophy of nineteenth-century science journalism: "Suppose it's Halley's Comet. Well, first, you have a half-page decoration showing the comet. . . . If you can work a pretty girl into the decoration, so much the better. If not, get some good nightmare idea like the inhabitants of Mars watching it pass. Then you want a quarter of a page of big type heads . . . and a two-column boxed freak containing a scientific opinion which nobody will understand, just to give it class."[1]

Today the style of science news is certainly more tempered, reflecting the norms of objectivity guiding the profession as well as the backgrounds and biases of reporters. But contemporary science journalism still bears the weight of its tradition.[2] The early efforts to communicate science to the public defined a role for science journalists, created expectations about their relationship to the scientific community, and shaped the attitudes of an emerging profession. To understand the present-day style of science journalism, we must first consider patterns and precedents established many years ago.

The Development of a Style

Nineteenth-century science and technology appeared in the press in both serious and sensational form, some of which would be nearly inconceivable today. In the 1830s the *Athenaeum,* a London newspaper, published pages on the regular meetings of the Geological Society of London. Both European and American newspapers published the complete lectures of such leading scientists as Thomas Huxley, Louis Agassiz, and Asa Gray, who traveled around to popularize their work. A special edition of the *New York Tribune* in 1872 published John Tyndall's physics lectures; it sold more than 50,000 copies.[3]

Most science journalism in the nineteenth century, however, consisted either of directly practical information about new farming techniques, the latest home remedies, or wildly lurid stories. This was the heyday of science hoaxes. In 1835 the press reported that astronomer Sir John Herschel had observed batlike human beings on the moon. A story in 1844 reported a three-day crossing of the Atlantic in a balloon. The newspapers published reports of scientific oddities and weird claims worthy of today's *National Enquirer:* the earth was flattening out; a man with a stomach could live without eating; and a woman with no stomach could eat without digesting.[4]

Later in the century, news reports were influenced by both the dawning awareness of the power of science and technology and the deep ambivalence toward the industrial revolution. Science seemed increasingly fascinating but obscure, benevolent but somewhat dangerous. Popular magazines began portraying science in nearly mystical terms as an awesome and remote activity performed by omniscient individuals. This, after all, was the century of Frankenstein and of Dr. Jekyll and Mr. Hyde.

The role of science during World War I (in particular Germany's exploitation of chemical research to manufacture explosives), together with the postwar proliferation of consumer goods, increased the public's awareness of the social and economic power of science. In words that sound familiar today, the press expressed concern about America's role in international competition in light of the German monopoly on patents in chemistry. A 1917 article in the *Saturday Evening Post* asked, "What's the matter with American chemistry anyway?"[5]

As scientific research expanded after World War I, increased public interest in science was reflected in a growing popular science press. This press focused mainly on applications; science became a way to get things done. According to Frederick Lewis Allen, "The prestige of science at this time was colossal. The man in the street and the woman in the kitchen, confronted on every hand with new machines and devices which they owed to the laboratory, were ready to believe that science could accomplish almost anything."[6]

Ironically, enthusiasm about science in the opening decades of the twentieth century also increased antiscience tendencies: witness the revival of astrology and mysticism and the antievolution activities of religious fundamentalists, who saw Darwinism as a threat to their values. The press was a vehicle for such antiscience views.

An important development—and a common theme in the press at this time—was the widening gap in knowledge between the scientific expert and the layman. In 1919 the *New York Times* published a series of editorials on the public's incomprehension of new developments in physics and the disturbing implications for democracy when important intellectual achievements are understood by only a handful of people. Einstein's theory of relativity became a symbol of obscurity. Morris Cohen, a friend of Einstein, wrote of the dilemma in a statement to the *Times:* "Free civilization means that everyone's reason is competent to explore the facts of nature for himself, but the recent development of science, involving ever greater mastery of complex techniques, means in effect a return to an artificial barrier between the uninitiated layman and the initiated expert."[7] This barrier, of course, separated journalists as well as the general public. But even as journalists were put off by the complexity of science, they were enamored of the progress it implied. They conveyed an image of science as an economic resource, an instrument of progress, a servant of technological needs.[8]

It was in this context that the newspaper magnate Edwin W. Scripps, founder of 30 newspapers and a syndicated press service, organized the Science Service in 1921. Science Service was the first syndicate for the distribution of news about science. Scripps believed that scientists were "so blamed wise and so packed full of knowledge . . . that they cannot comprehend why God has made nearly all the rest of mankind so infernally stupid." He believed that science was the basis of the democratic way of life. And above all he believed that, given the enormous social and technological changes of the period, science news would sell. With the help of a prominent zoologist, William E. Ritter, he engaged the cooperation of the National Academy of Sciences, the American Association for the Advancement of Science, and some leading journalists to translate science into "plain United States that the people can understand."[9]

Early in the formation of the science syndicate Scripps pondered a critical decision about its relation to scientific associations. Should the syndicate act as a press agent for the associations or as an independent news service? While hoping to avoid simply disseminating propaganda, he chose the former role. The syndicate was controlled by trustees from the most prominent science associations, and its editorial policies were dominated by the values of the scientific community.[10]

Presenting science in a marketable (that is, readable) form, the syndicate sold its articles to over 100 newspapers during the

1920s, reaching more than seven million readers—one-fifth of the total circulation of the American press. It laid the foundation for contemporary science journalism, giving the profession both a purpose and a style.

The Science Service's statement of purpose emphasized "the importance of scientific research to the prosperity of the nation and as a guide to sound thinking and living."[11] Ritter, however, stated the service's purpose from the perspective of scientists: he hoped that it would gain financial support for science and encourage the "mental attitude of science" among newspaper readers.

But selling science to the public, the founders concluded, meant making some compromises. Although anxious to avoid the popular style of yellow journalism, the first editor of Science Service, chemist Edwin E. Slosson, believed science writers had to compete for readers by catering to prevailing popular taste. Diagnosing this taste, he concluded that "It is not the rule but the exception to the rule that attracts public attention."[12] In presenting science to his readers, Slosson emphasized human interest: one advertisement for the service announced that "Drama and romance are interwoven with wondrous facts, helpful facts"; another that "Drama lurks in every test tube." Science Service articles cast science as a new frontier and scientists as pioneers and discoverers. "The pure thrill of primal discovery comes only to the explorer who first crosses the crest of the mountain range that divides the unknown from the known."[13] Scientists were in addition described as industrious, persistent, and independent; all in all, they incorporated the most positive values of American culture.

In keeping with its statement of purpose, the Science Service portrayed science not only as the basis for technological development and economic progress, but also as a guide to correct thinking and appropriate behavior. For example, it gave prominent coverage to eugenics—a prevailing belief among scientists of the day—on the grounds that the masses must understand "that the fate of the nation depends on how they combine their chromosomes."[14]

Slosson soon became recognized as the most influential interpreter of science to the public. His service, shaped by both perceptions of public taste and the values and concerns of the scientific community, created a market for science news and a pattern for the emerging profession of science journalism.

The dozen or so science writers who began their careers in the thirties perpetuated the ideas set forth by the Science Service and adopted a similar style. The best known of these writers were not formally trained in science, but they conceived of themselves as missionaries, committed to science at a time when few people cared. William Laurence, a science writer for the *New York Times,* described his profession: "True descendants of Prometheus, science writers take the fire from the scientific Olympus, the laboratories and the universities, and bring it down to the people." He believed that science was "a way to rise beyond the disappointments of the real world."[15] Waldemer Kaempffert, also working for the *Times,* wrote eloquently of his "boundless faith in science, a faith justified by past achievement." He envisioned engineers who would provide us with "thousands of small towns with plenty of garden space, low rents, breathing space."[16] Gobind Lal of the Hearst newspapers defined his role as a science writer in similarly optimistic terms: "My job is to create a public taste for science. We must make science accessible to the people and for the people."[17] These writers became the gurus of science journalism when it burgeoned as a distinct specialty with the rapid development of science and technology in the years following World War II. They expressed a view of science and technology and developed a style that continues to be reflected in the heroic images of science that mark science writing today.

Norms of Objectivity

Contemporary science reporting reflects early efforts to adapt the norms of scientific objectivity to the practice of journalism.[18] Journalists no longer believe that real objectivity is possible, but they are expected to approach the ideal of neutrality and unbiased reporting by balancing diverse points of view, by presenting all sides fairly, and by maintaining a clear distinction between news reporting and editorial opinion.

The Society of Professional Journalists (Sigma Delta Chi) has formalized these norms in its code of ethics, which requires journalists to perform with "intelligence, objectivity, accuracy, and fairness" and to free themselves from all obligations, favors, or activities that could compromise their integrity.[19] Accordingly, many newspapers place constraints on journalists to assure their neutrality. Reporters on the *Washington Post,* for example, must avoid involvement in politics or community affairs that could compromise or seem to compromise their ability to report with fairness.[20] The *Wall Street Journal* prohibits news employees from receiving lecture fees from profit-making firms. The *New York Times* prohibits its reporters from receiving fees from groups whose "interests they cover." A Sigma Delta Chi survey of 900 newspaper executives found that half of the editors would not allow reporters to accept free trips under any circumstances. But journalists are increasingly on the lecture circuit, and big-name reporters earn large fees, raising uncomfortable questions about the actual meaning of objectivity in the news.

Objectivity in the press is an American ideal; European newspapers are expected to have an explicitly partisan view that is understood by their readers. In *Objectivity and the News* (1981), Dan Schiller attributed our ideal of objectivity to three related factors.[21] In part it developed as a reaction against the excesses of yellow journalism in the nineteenth century. Objectivity also became economically important with the organization of the

Associated Press in 1848. This centralized newsgathering service had to sell its articles to newspaper clients with quite diverse political views, so a neutral style of reporting was required. A third explanation can be found in the growing influence of the scientific attitude in the nineteenth century. Central to this attitude was the belief that "facts," standing high above the distorting influence of interests and pressures, can and should be distinguished from values. The press, in effect, adopted the ideals of science at a time when science was becoming broadly accepted as an apolitical basis of public policy, a model for rationality in public affairs.

The links between the ideals of science and the norms of journalism began to be forged during the 1830s. Prior to this time the openly partisan "party press" had dominated newspaper publishing. With the appearance of the "penny press" in the early 1830s, the character of American newspapers began to change. Depending on commercial advertising rather than political patronage, the penny press adopted norms of objectivity on the assumption that factual reporting would promote understanding and enhance democracy. In May 1835 James Gordon Bennett, editor of the *New York Herald,* promised: "We shall endeavor to record facts on every public and proper subject, stripped of verbiage and coloring." Later, in 1840, he expressed the purpose of journalism: "I feel myself in this land to be engaged in a great cause—the cause of truth, public faith and science against falsehood, fraud and ignorance."[22]

The editors of the penny press believed that science, with its assumed reverence for facts, was the guide to proper journalism. In a society strained by factional disputes, expressions of ideology and political views were to be avoided. A poem in the *Philadelphia Public Ledger* in 1839 stated the goal of the emerging newspapers:

We strike for right, and will not spare a blow,
In bold defiance still our ensign show;
All cliques, all sects, all parties we despise,

Above all partial motives proudly rise;
The laws our guide, the good of all our aim,
We yield no principles for transient fame.[23]

The norms of objectivity in journalism were reinforced throughout the nineteenth century to encourage the values of pluralism and promote a democratic process based on equal public access to "facts." An 1884 handbook for journalists, for example, states the imperative of separating facts and values in reporting and relates this imperative to American democratic values: "It is as harmful to mix the two in journalism as it is to combine church and state in government."[24]

There were, of course, cynics; as Mark Twain put it: "First you get your facts, then you can distort them as much as you like." However, by the turn of the century belief in science as the embodiment of neutrality and rationality was firmly entrenched. Thus, scientific values penetrated many social and political institutions: witness the increased emphasis on technical expertise in government, the growth of realism in literature and art, and the political reforms of the progressive movement. In this context a scientific—that is, neutral—presentation of the facts was defined as the enlightened basis of a responsible press.

This spirit of objectivity converged with the growing idea that scientific knowledge prevailed over all other forms of knowledge to shape the conventions of journalism. In 1915, for example, the *New Republic* proposed that journalism schools seek to create "a morale as disinterested and as interesting as that of scientists who are the reporters of natural phenomena. News gathering cannot perhaps be as accurate as chemical research, but it can be undertaken in the same spirit."[25]

Following World War I, political propaganda and the activities of the growing profession of public relations led to an awareness of how facts could be manipulated and, inevitably, to a certain skepticism about the reality of the "value free" press. Indeed, faith in the objectivity of facts was soon considered naïve. Newspapers responded by adopting various practices to prevent the manipulation of information, including the use of bylines, the careful identification of news sources, and an increase in explicitly interpretive reports. However, objectivity remained an ideal. It served the same purpose for journalists as it did for scientists, helping both professions maintain autonomy and independence from public control. Just as scientific objectivity enhanced the legitimacy of science, so it enhanced the legitimacy of journalists, allowing them access to sensitive information and justifying their protection under the First Amendment.[26]

The ideal of objectivity did not limit the seldom qualified enthusiasm in the press for scientific, medical, and technical advances. Immediately after World War II (and even during it), journalists wrote of the "promises" of peacetime applications of atomic energy, the "progress" in aviation, the "revolutionary developments" in pesticides, vaccines, and drugs. Above all, they hailed the "cosmic breakthroughs" of the space program.[27] Because of the temper of the times, discussions of relative values, of risk, of objectivity, were suspended; neither journalists nor for that matter scientists gave much consideration to the social implications or even the cost of science and technology except in the Cold War context, when weapons were involved.

Not until the 1960s were significant doubts cast on the meaning of objectivity as a guide to journalism. Reflecting the political mistrust of the period, 1960s media analysts suspected that the "objective" press was simply in collusion with institutions of power; objectivity was viewed merely as a mystification and as a convenient myth. Journalists' efforts to maintain objectivity, according to sociologist Gaye Tuchman, were only a "strategic ritual"; subjective perceptions inevitably entered their writing, while the tone of objectivity allowed reporters to avoid responsibility for their views.[28] Similarily, Todd Gitlin argued that norms of objectivity only reinforced the "dominant hegemony"; by conveying an image of neutrality, journalists in fact strengthened the existing legacy of private control over production.[29] Harvey Molotch and Marilyn Lester denied even the possibility of objectivity; all news is manufactured, for there is no "world out there" to be objective about.[30]

Journalists themselves became more aware of how their own values entered their writing. As one experienced science writer told me: "A good news story is filtered through values. In the course of writing about science, I absorb an enormous amount of fact and information. I have to sort it out. My ideas about the role of the public and private sphere, my values concerning the environment, all help me to organize my material. We understand that. Scientists often fail to understand; yet they, too, carry baggage."[31]

The idea that standards of scientific objectivity can be met by fair and balanced presentation of different points of view persists, however, as is evident in the reporting of technological controversies. Journalists try to maintain balance by quoting scientific sources representing opposite sides of a dispute whether it be over toxic wastes, artificial sweeteners, or biotechnology. As a reporter who writes on controversies put it, "As long as you don't fall into the trap of presenting just one side, you're playing ball." Ironically, this notion of objectivity is meaningless in the scientific community, where the values of "fairness," "balance," or "equal time" are not relevant to the understanding of nature. On the contrary, scientific standards of objectivity require not balance but empirical verification of opposing hypotheses. Simply to balance sides gives readers little guidance about the scientific significance of different views. Though journalists' norms of objectivity were initially modeled on scientific method, their current implementation in reports of scientific disputes is very often a source of irritation to the scientists involved.

Changing Professional Ideals

In the 1960s the expansion of an advocacy press with a critical and reformist ideology forced the mainstream press to reconsider the conventions of journalism, and news articles became more interpretive, investigative, and adversarial in character. Science writing reflected these trends. Moreover, as scientists themselves entered as advocates on various sides of the growing number of environmental and energy disputes, the political dimensions of science became news, and the role of scientists as sources of objective and neutral information came increasingly into question.

During this time a number of science journalists began to criticize the "missionary" role of their own profession. In 1966 Henry Pierce of the *Pittsburgh Post Gazette,* for example, called science journalists "a bunch of patsies prone to uncritical acceptance of anything we are told by our authorities—our authorities being doctors and scientists." He observed that other journalists maintained a more healthy skepticism toward news sources: "But we, bless us, go in with our bright baby-blue pencils poised, faithfully recording anything our scientists—gods—tell us. Never does it occur to us that these guys too may have motives that are less than noble."[32] John Lear, editor of the *Saturday Review,* wrote, "The spirit of untrammeled inquiry and skepticism required of journalism in other fields must become a standard in science writing."[33] And David Perlman chided his colleagues: "We are in the business to report on the activities in the house of science, not to protect it, just as political writers report on politics and politicians."[34]

Breaking away from the constraints of neutrality, some reporters during the 1970s captured the spirit of social criticism by sympathizing with the questions provoking environmental controversies and the growing concern about the social impact of science and technology. They tried to become critical investigators rather than passive conveyers of scientific and technical information.[35] They began, for example, to write more about the social implications of new technological developments. They increasingly reported disputes: over the antiballistic missile (ABM), the supersonic transport (SST), food additives, nuclear power, and environmental pollution. And they became increasingly sensitive to the implications of science and technology for such diverse areas as business, medicine, environment, energy, and even crime control.

In the 1980s, science writing, like other areas of journalism, returned to a less critical and more promotional style—except in the face of disasters. Journalism students in this image-conscious decade focused their studies more on techniques of public relations than on reporting and editing skills. Reports on science and technology were adapted to the conservative business mentality of the Reagan administration, and, reflecting the heritage of the science writing profession, images of heroes, breakthroughs, and frontiers again filled the news. The 1990s has revived a more critical style of science reporting. The end of the Cold War brought a search for new and newsworthy villains, scientists among them. Incidents of scientific misconduct, the costs of megascience, and bioethical dilemmas are sure to attract current media attention.

These general changes in the reporting of science and technology can be seen even in the careers of individual journalists. Consider, for example, David Perlman of the *San Francisco Chronicle,* one of the most experienced and prolific science journalists in the United States. A former president of the National Association of Science Writers and winner of several journalism awards, he is also one of the most highly respected writers within the profession. His career has spanned the post–World War II period of dramatic scientific and technological growth.

Perlman earned a degree in journalism in 1940 with no special focus in science. After serving in the army during World War II, he stayed on in Europe as Paris correspondent for the

New York Herald Tribune. He came to the *Chronicle* in 1951 as a general assignment reporter and began to write mainly about environmental issues. He was attracted to science in 1958, when a friend gave him Fred Hoyle's *Nature of the Universe,* and began to specialize in this area of journalism. Science was local news in the Bay Area, with its concentration of research universities and affiliated laboratories. Just by focusing on the "local" research at Stanford, Berkeley, and the Lawrence Livermore Laboratory, Perlman could essentially cover the most important areas of contemporary science and the latest technological developments. In addition, Perlman has traveled all over the world, covering professional society meetings, space shots, planetary probes, and scientific expeditions.

An examination of Perlman's articles in three critical years, 1960, 1972, and 1982, reveals important shifts in orientation toward science and technology that are characteristic of many reporters.[36] His writing in 1960 on science and technology expressed both his enthusiasm and awe as a new science writer and the general post-Sputnik optimism. Many articles were about advances in medical research. His language was extravagant. Medical researchers were detectives and wizards seeking "clues," "probing secret structures," "unlocking stubborn secrets," operating with "flashes of insight." They were also fighters "mobilized into assault teams attacking disease," and they needed public support.

Computers, military and space technology, research instrumentation, and especially nuclear power were described with similar enthusiasm as "super systems." The rocket was "a power-boost in a race to space"; a computer was "a damned machine that talks back with a flicker of ten red eyes." The peaceful atom was "the key to the future" in the "atomic horse race." Perlman wrote about the use of atomic energy for atomic flight, cheap fuel, monitoring smog, space exploration, deep sea mining, and fish farming.

The hazards of technology—especially radioactivity and atomic waste, later to become a critical theme in his writing—

received some attention. Yet in one article he welcomed the development of a new chemical pesticide as a boon and a break-through, with no hint that it could be hazardous, although it was derived from poison gas. The articles on hazards conveyed few of the concerns that were to dominate Perlman's writing a decade later. Rather, they are reassuring reports of agreements designed to protect the public, suggesting, for example, that "we are keeping tabs on atomic garbage."

Paralleling changes in the public and journalistic views of science and technology, the tone and style of Perlman's writing shifted rather remarkably by 1972. There were no reports of breakthroughs, far less drama, and many more qualifications. His articles on medical science—even those on organ transplanta-tion—insisted on the limits of scientific claims. Perlman also wrote critically about the potential legal and ethical problems of genetic research, the politics of sickle-cell anemia research, and the controversies over acupuncture. The articles on advanced technology remained enthusiastic, but he reminded his readers of mounting public hostility to costly projects such as the further exploration of space. He wrote on the hazards from oil spills, food additives, pesticides, nuclear power, and air and water pollution, always emphasizing the controversies surrounding these issues. Describing the growing number of disputes among scientists about "troublesome technology," he observed the "tumultuous hazards" of scientists' involvement in public issues.

By 1982 Perlman was expressing the renewed technological optimism of this period. His language was more inflated; in writing of medical research he noted the "revolutionary" advances at the "frontier" of genetics and "promising" new diagnostic tech-niques. In covering biotechnology he employed an aggressive language: "fast breaking competition," "explosive growth," "revolution," "turmoil." In covering basic research he wrote about "cosmic mysteries," "historic findings," "adventures," and curiosi-ties (a fish that likes to get hot). Later in the 1980s, Perlman's reporting reflected the expanded public interest in ethical and value questions. He wrote extenively on human rights, ethics, and secrecy in science.

Perlman and other science journalists adapt their writing to the spirit of their times and use their personal instinct of what readers and "opinion leaders" will find of interest. As one reporter told me, "I consider myself the average general public. I think that if it is interesting to me, hopefully it will be interesting to the people I write for."

Social Biases of Science Writers

Journalists' selection of newsworthy issues and their tone in writing about them are products not only of their professional styles but also of their educational background and social biases. The early generation of science journalists were seldom trained as scientists. Perlman, who describes himself as typical of science writers over 50, claims that he hated science in college (Columbia University) and took only the minimum requirements to graduate. "I started out, wearing a snapbrim fedora on the back of my head, covering fires; wore a trench coat in Europe as a correspondent for the *Herald Tribune,* and when I first started covering science, I truly thought that bilirubin was the name of the patient a doctor was describing when she talked about advances in treating hepatitis! . . . When *Sputnik* went up, I couldn't even have told you why it didn't fall right back down. Everything else I've learned on the job—out of enthusiasm, covering stories, reading a lot of background."[37]

Walter Sullivan, another veteran science reporter, joined the staff of the *New York Times* as a music critic. World War II turned him into a foreign correspondent, and then a series of trips to the Antarctic in 1956 transformed him into a science writer special-izing in geology, astronomy, and physics. He and his professional peers developed familiarity with science mostly on the job through expeditions, press conferences, and occasional seminars. Some journalists took on temporary scientific jobs or internships, acting, in effect, as participant observers. One spent two weeks with a scientist tagging turtles in Florida; another joined an oceanographic research vessel for three weeks; a third went on an expedition to the Galapagos Islands. Others have worked in various research laboratories, learning the language and methods of science. Younger science journalists, however, have far more formal science background, often having majored in science as undergraduates. Some reporters for major newspapers began their careers as writers at magazines such as *Science* and *Nature*. And some have gone through special science journalism graduate school programs, which include a science curriculum.

But most journalists who cover science and technology, especially those working for small-town newspapers, write about science only part of the time. And even general reporters, when covering national security, crime, trends in education, budget priorities, or health, must often touch on some scientific or technical issues. These generalists often find the science beat confusing. Afraid of technical complexity, they are apt to avoid substantive questions. And lacking both training and experience, they are often unable to evaluate what they are told.

Science journalists themselves are divided as to the importance of formal training in science. William Stockton of the *New York Times* argues that science-trained journalists can be more critical about shoddy research methods and are less likely to take what they're told at face value than their untrained colleagues.[38] Reporters who know too little about science may not know how to find technical resources, what questions to ask about a complex matter, or exactly what to make of the answers. Preoccupied with reaching a basic understanding, they have little time or energy to interpret underlying issues.

Numerous cases support the need for greater methodological sophistication among journalists. For example, during a press conference in which a scientist described his cancer research, a probing and experienced science reporter found out that inbred rats had been used in the experiments—an error in the selection of sample research subjects that called into question the validity of the results. He refused to cover the story. A less experienced reporter, however, wrote it up without comment, having failed to understand that the methodological flaw undermined the significance of the findings.[39]

While agreeing on the need for greater technical sophistication, some journalists argue that too much science education can handicap the reporter. An experienced reporter without a background in science but with a generally critical eye can use journalistic skills to force scientists to explain things carefully. According to one journalist, "The generalist can often get away with asking stupid questions better than the science-trained journalists." He admitted his quiet appreciation when a less specialized colleague asked naïve questions that helped to clarify a complex issue. Moreover, if a journalist knows too much about a technical subject, his writing may become overspecialized.

More important, journalists trained extensively in science may adopt the values of scientists and lose their ability to be critical. Journalists worry about this: in interviews they suggest that "the science writer who is basically a scientist looks at things in terms of what it means to the development of science, but the journalist looks at it in terms of how it is going to affect people and their quality of life." According to some reporters, "Science writers who are scientists have more reverence for scientists than do other journalists. . . . They sometimes draw back from tough questions because of the very deep respect they have for fellow scientists." Others, however, contend that it is the least experienced reporter who is the most reverent, that science-trained reporters are "more jaded and therefore less vulnerable."[40]

Whatever the effect of technical background, political and social biases clearly influence the attitudes of science journalists and their choice of newsworthy events. This is most evident in observing what issues fail to appear in articles on major aspects of science and technology. For example, with all the stories on chemical waste and chemical carcinogens, few journalists have written about risks to workers in the chemical industry. As intrusive neighbors, chemical companies are newsworthy; as employers, they are not. Even the extensive coverage of the lawsuits generated by workers exposed to asbestos focused more on bankruptcy manipulations by the Johns Mansville Corporation than on the health conditions of its workers.

Science writer Paul Brodeur speculates on the reasons for the remarkable dearth of articles on this subject. "I submit that if a million people in the so-called middle or professional class were dying each decade of preventable ocupational disease . . . there would long ago have been a hue and cry for remedial action."[41] David Burnham, science writer for the *New York Times,* explains why most reporters avoid the subject of occupational health: "A class bias exists and I have to calculate my stories somewhat to that reality. The upper-middle-class people who read the *Times* can shrug their shoulders and see cotton dust as very distant, as a minor problem affecting some poor textile worker down south. But they can identify with a story of environmental carcinogens. Cancer terrifies everyone."[42] Likewise, Wade Roberts of the *Texas Observer* (an advocacy newspaper) confirms that "while the classes that own the newspapers are sometimes willing to take a strong stand on environmental health, occupational health just doesn't interest them."[43]

When journalists do cover occupational health issues, they go to official sources rather than to workers. In 1976 a dramatic incident took place at Electric Boat, a shipyard in Groton, Connecticut: it was announced that 1200 workers were found to have traces of asbestosis. This generated 50 stories in local newspapers, but only one reporter interviewed the afflicted asbestos workers themselves.[44] Karen Rothmeyer of the *Wall Street Journal* explains this bias bluntly: "Middle class journalists who are used to dealing with middle class officials won't get off their asses to make the difficult effort to find people on the other side. . . . Too many reporters wind up being Establishment stooges, not because they're uncaring people, but because they're middle class and don't want to struggle with speaking another language with different people."[45]

Class biases are reinforced by political preferences, or even the lack thereof. Journalists, contrary to the views of many critics, who see them as belonging to the liberal left, tend to be apolitical or middle-of-the-road. Herbert Gans, a sociologist who has studied journalists, characterizes them as "members of the extreme center . . . suspicious of the oratory, skeptical of grand plans, committed to rational programs to solve problems."[46] The profession, devoted to the notion of objectivity, discourages those with strong political convictions. Politicized writers prefer to work for the alternative or advocacy press. Even journalists who cover numerous risk controversies for the daily papers seldom question the "system"; they prefer to conceptualize each problem separately in terms of aberrant individuals, bad companies, or simply ignorance.

Science journalists generally try to avoid expressing a political view. Like scientists, they want to separate science from politics. They rely extensively on scientific publications such as *Science, Nature,* and the *New England Journal of Medicine* and feel an obligation to report on the latest scientific findings as they appear in these sources. They define their role more as explaining science than as analyzing its politics. When questioned about the politics of science, they will argue: "If I deal with the politics of an issue, then I stop being a science writer," or "I want good science, not moralizing."

The assumption that politics is incompatible with science, however, can lead to conflicts of interest that undermine norms of journalistic neutrality. In 1979 the Associated Press sold a series of Alton Blakeslee's articles on cancer research to 230 newspapers. The AP failed to note in the articles that Blakeslee was a paid consultant to the American Cancer Society and had originally written the series for them. In light of the heated politics of cancer research, Blakeslee and others questioned the ethics of this oversight. But the AP editor could not understand their concern: "That's like saying that God is political."[47]

This lack of political perspective, however, fosters uncritical trust in scientific sources of information. Journalists rely on information provided by scientists, even in highly commercialized areas like biotechnology where economic stakes are bound to shape their sources' views.[48] Attention turns to conflicts of interest only in response to scandals or disputes.

The reverential attitude of most science journalists toward their subject further reduces professional skepticism. David Perlman sums up the vision: "The science writers' words have captured the nobility to which the human mind can rise when it turns from violence and bigotry to an exploration of the infinite order of the universe. Political writers grow cynical, city editors grow old; only science writers stay young and endlessly excited and appreciative of elegance. For science by definition is young and exciting and elegant."[49]

Captivated by science and regarding scientists with awe, most science journalists write about their subject in much the same glowing terms that sports reporters use for prominent sports stars. The veteran science reporters who have been on the beat since the initial space launches at Cape Canaveral are especially eloquent about science and technology. This influential group of writers, having spent long hours together waiting for space shots, developed a lasting camaraderie. During that dramatic decade they gained an overwhelming sense of the excitement of science and technology; they felt that they were a part of the action.

Perlman, for example, describes "the excitement of witnessing scientific achievement firsthand; to watch the data on Jovian radiation pour in from *Pioneer 10* in its flyby past Jupiter; to sit with Arthur Kornberg in his laboratory as he draws two ring-shaped helices in his synthetic viral DNA."[50] Perlman is not alone. Science, to Victor McElheny, formerly of the *New York Times,* is "the Icarus." Describing his role as "personal witness to a golden age of scientific exploration," he writes: "To be present in the Lunar Receiving Laboratory outside Houston, Texas, when the first box of igneous rock and dust from the moon was opened; at the Jet Propulsion Laboratory in Pasadena when the first vertical strips of the first panoramic photograph from an automated *Viking* station on the planet Mars appears on the screen—a romantic observer could liken such opportunities to being a spectator at Barcelona when Ferdinand and Isabella stood to receive Christopher Columbus on his first return from the New World, or at Deptford when Queen Elizabeth came down to see Francis Drake, back home from circumnavigating the globe, and dubbed him knight."[51]

As the veteran science writers retire, the younger generation of journalists are more critical. Reflecting broader social trends, they are more conscious of the social, ethical, and economic costs of science and technology. But they too find their subject endlessly entertaining. Their language of magic and wonder ("it boggles the mind") expresses their pleasure in describing science and technology. Many of them are amateur science buffs, endlessly curious about scientific issues. "I'd love to be a scientist," one journalist told me. "When I work on a story I get to sit at the feet of the most luminous minds in the U.S. It's more fun than anything I've ever done," said another. At the same time science writers often feel insecure when they interview scientists: "I feel naked without credentials."

This attitude of awe and admiration differentiates most science writers from political reporters, who are much more inclined to look critically at the events that they cover. Some, however, maintain a certain distance and skepticism. Ed Edelson of the *New York Daily News* declares (only partly tongue-in-cheek): "I refuse to believe that my purpose is to keep the public informed about the true nature of scientific research. If I truly did tell people about most scientific research, it would mean an end to public funding of that occupation."[52]

Most science reporters tend to behave rather like sports writers: they have chosen their topic out of love for it. They are driven, according to one writer, by "inner-directed ambition"; according to another, by "a sense of wonder and awe." While general reporters try to maintain a certain detachment from their subject by rotating beats, most science writers (like sports writers) remain with their specialty for years. Discussing their work, reporters say: "I cannot think of any job I would rather have"; "It is constant change, constant excitement"; "We're in it because we love science and it is a job in which you can keep learning more and more about science all the time"; "Science has made some wonderful advances; of course we view it as positive"; "We want to sell science."[53] The more sophisticated writers may make fun of the trite language of "breakthroughs," but the background and biases of science journalists, as well as the precedents of their profession, perpetuate this style of reporting. Perhaps more important, their biases lead them to identify more closely with their subject and their sources than do journalists in most other fields. This identification, however, has placed them in an awkward position—between two professions with quite different expectations. They strive to maintain the respect of their scientific sources and to satisfy the ideals of science, but they must, first and finally, meet the constraints of their own profession.

7

CONSTRAINTS OF THE JOURNALISTIC TRADE

At the end of 1982 the Centers for Disease Control (CDC) reported 800 cases and 350 deaths from AIDS, clearly indicating that the nation was facing a serious public health problem. However, except for the gay press and the *San Francisco Chronicle,* influenced by the political clout of the gay community in that city, the media virtually ignored the problem until May 1983.[1] It was not that reporters were unaware of the issue. Lawrence Altman, M.D., of the *New York Times,* wrote an article on AIDS in 1981 that was not published; in 1982, Jerry Bishop's editor at the *Wall Street Journal* also turned down an article. In contrast, the *San Francisco Chronicle* hired a full-time AIDS reporter, Randy Shilts, that same year.[2]

Quite abruptly, in May 1983, news coverage of AIDS expanded, mainly because of an editorial written in the *Journal of the American Medical Association (JAMA)* by Dr. Anthony Fauci, director of the National Institute of Allergy and Infectious Diseases, who raised the possibility that AIDS might be transmissible to the entire population through "routine close contact." To the media this suggested that everyone was at risk. Around the same time, studies began to suggest that there could be heterosexual transmission. Defined as a homosexual disease, AIDS had attracted little public attention: when it seemed that it might extend beyond the gay community, coverage significantly expanded.

Following reassurance on this point from scientists, the coverage temporarily declined after 1983. From the lack of media interest, one might have concluded that AIDS was not a very important issue, although this was a period of critical scientific advances and growing awareness of the seriousness of the disease and its implications. Media interest picked up in the summer of 1985 with the illness of Rock Hudson and announcements of epidemiological studies indicating the exponential spread of the disease. In June 1985, the *New York Times* only ran four articles on AIDS; there were sixteen in July, 46 in August, and 72 in September. At the end of that year, the *New York Times* financed Lawrence Altman's long series of articles on AIDS in Africa—a rare example of well-funded, detailed, investigative science reporting.

After 1986 major newspapers began to cover AIDS, often providing solid and accurate technical information as well as personal dramas. In fact, AIDS reporting, assigned to the most experienced medical journalists, is technically detailed despite the many scientific uncertainties about the etiology of the disease. Yet, shaped by sexual conservatism and reflecting the moralistic stance of many governmental authorities, news reports often convey an unrealistic and even counterproductive social message about how to prevent the spread of AIDS—abstain.

The conservatism of the media in reporting on AIDS has reflected editorial concerns about alienating readers. True, things had already changed by the late 1980s. Earlier in that decade, the *New York Times* had refused to print the word "gay" except in a quoted passage. By 1986, Surgeon General Koop and columnist Jane Brody were giving televised instructions on how to use condoms. Yet moral judgments about homosexuality continued to shape AIDS coverage as article after article appeared on homosexual promiscuity. The press labelled AIDS not as a viral disease like hepatitis, but as a "sexually transmitted disease" like syphilis, clearly laying the blame on immorality. The STD label helped to stigmatize those with AIDS. Searching for a cause, journalists used a language of blame and rebuke.

Such views, of course, did not originate with the press. Public health departments were also equating AIDS with STDs and even the gay press was linking the "fast track" life style to the disease. But journalists, ideally an independent voice, provided little critical analysis that might have called early attention to the growing number of IV drug users and women and children with AIDS.

A further dimension of AIDS coverage has been its remarkable polarization. On the one hand sensational stories and headlines warned of the vulnerability of all, making Fauci regret his early speculation. A *Life* story announced: "AIDS breaks out of high risk groups": "No one is safe from AIDS." A *Time* magazine headline read, "A scourge spreads panic." Metaphors comparing AIDS to leprosy, the plague, or a time bomb repeatedly appeared. Headlines called attention to the "deadly new epidemic," "the public health threat of the century." "Is there death after sex?" they asked: "Even you can become infected." On the other hand reports reassured the public by framing AIDS in terms of "high risk groups." AIDS became a disease for "others": gays, drug users, Africans, Haitians, those who are somehow immoral. This, in effect, ghettoized the disease by defining it only as a problem of those engaged in particular life styles—*their* problem, perhaps affecting their spouses and children, but not a disease of society.[3] The fact that AIDS has had a profound impact on American culture and institutions—health care, social services, prisons, the family, the role of government—has been systematically underplayed.

In the polarized coverage of AIDS, journalists were responding to their sources of information, and public officials tried to play down the AIDS risk in order to avoid panic. But advocates from the gay community wanted dramatic stories about the dangers of an epidemic as a means to gain much-needed resources for research and medical care. And conservatives encouraged the tone of reprobation as a way to further their moral agendas. AIDS reporting exposes journalists to strongly conflicting pressures and sometimes bitter reproach. For example, in 1990, when Gina Kolata of the *New York Times* wrote an article criticizing the early distribution of a risky new AIDS therapy, AIDS activists responded with a scathing and vehement denunciation. And in 1994 when journalists covering preliminary trials of an AIDS vaccine reported that five volunteers became infected, scientists reproached them for compounding the difficulties of launching full-scale clinical trials.

Articles on AIDS activism have proliferated in the 1990s. The press is attracted to the growing conflicts—over condom distribution, sex education, free clean needles, and HIV testing. The dispute over testing in hospitals becomes, in the press, a conflict between physician autonomy and patients' rights, and the emphasis is on irreconcilable values. The press has paid little attention, however, to the deeper social tensions revealed by AIDS—over society's commitment to individual autonomy when community values are at stake, over the roles and responsibilities of government in managing disease, over tradeoffs between scientific research and other costly programs, and—until 1994—over the appropriate organization of a humane health care system.

This brief history of the media coverage of AIDS illustrates the constraints under which journalists, especially science journalists, work. Writing about this disease required the time and budget to cover a complex subject and to develop interpretations in the face of technical uncertainties and scientific disagreements.

Journalists sought angles that would hold the attention of readers and appeal to editors. They had to accommodate public sensitivities and social biases. And they were vulnerable to sources of information with conflicting agendas. These constraints, implemented through editorial policies and practices, affect the work of journalists in all fields, but they pose special problems for those reporting on the complex, uncertain, and often slowly evolving events that characterize many aspects of science news.

Newswork

Science journalists obtain material for their stories from press releases, public relations officers, professional society meetings, press conferences, scientific journals, and interviews. They usually write their stories rapidly, pressing to meet their deadlines and to beat their rivals on other newspapers. The competitive quest for dramatic stories affects the pace of daily newswork, encouraging a focus on "breaking news" and discouraging the coverage of long-term issues or issues that require extensive technical background.[4] "It's the constraint of getting out daily stories," reporters told me. "We can't take time off to do something that's not breaking news. We go with something that is there."

Emphasis on breaking news is often detrimental to good coverage of science, for important issues such as AIDS may not be associated with striking single events, and significance usually lies in long-term consequences. In the case of dramatic new surgical techniques, their significance rests on the patients' postoperative histories. Yet media attention tends to wane after the initial dramatic event, unless, as in the case of Barney Clark, repeated traumas keep the issue—if not the patient—alive.

Similarly, the focus on "breaking news" limits analysis of the methods and processes of science; however important in assessing the significance of research, methods are not considered news.

Experienced science writers understand the importance of research methods; they often treat press conferences much like a graduate seminar, questioning research procedures and experimental controls. But when they actually write up a piece of research they tend to focus on new findings or discrete events.

Because breaking news is scooped by the daily press, writers for weekly magazines are more inclined to provide background material and to write reports on scientific or technological trends. For most magazine editors, however, science is a low priority. A *Newsweek* reporter identifies the "Friday Syndrome": "Every Monday, editors decide to do special pieces on science . . . by Friday, a disaster will take place or Congress does something controversial and our editors push these pieces off the seat."[5]

Science newswork is also constrained by limited budgets. Most papers under 100,000 circulation employ only general reporters who cover several beats, because engaging a specialized reporter to cover science is so costly. Only the largest papers are likely to have separate specialists for, say, environment, medicine, and energy. That the *San Francisco Chronicle* had a special AIDS reporter was unique. Similarly, few papers allocate significant allowances to science reporters for travel to meetings or interviews. While the *New York Times* provides generous budgets for leading writers like Lawrence Altman, most science reporters must compete with their colleagues in other fields for funds.

Science writers also compete with political writers for space and complain of problems convincing editors that science is as newsworthy. Journalist Earl Ubell once described the struggle of the science writer who has an ongoing, important story: "Frequently the first time he writes a story it appears somewhere behind the financial pages in the newspaper. The second time he writes the same story, with approximately the same lead, it moves forward so that it might get on the second page of the paper. And, finally, the third time around it almost always hits the front page because by that time the editor has begun to understand that perhaps it is news and that he ought to do something about it."[6] The amount of space allocated to science has increased over the past few decades, but with few exceptions, competition for space limits the possibility of including the background material and qualifications useful in conveying complex technical issues. Problems of space were eased in the 1980s by the introduction of science sections in many newspapers. For a while these proliferated: there were 19 in 1984, 66 in 1986, and 95 in 1989. But by 1992, more than half of these were gone and the others reduced in size.[7]

Deadlines present a further constraint. Shortage of time limits the number of sources that a reporter can use. According to a study by media analyst Sharon Dunwoody, half of the articles produced at a scientific meeting by reporters who had daily deadlines cited only a single source.[8] Those sources who can provide information efficiently in a form that is predictable and easy to turn into a story—usually experienced public relations people—are most likely to influence the shape of the news.

In recent years, the use of computers has eased the time constraints of journalism. Reporters with access to data banks can gather information more rapidly, but the demands of daily story production still encourage the use of prepackaged information that has been organized specifically for the press. Reliance on such preconstituted accounts as press conferences, news releases, and computer-controlled information is, of course, no substitute for personal investigation, but it is a practical and therefore popular means of getting a job done under the pressure and routines of daily newswork.

Editorial Constraints

Journalists constantly talk about "headline idiocies and the ignorance of editors,"[9] for editorial judgments mediate their work. While editors play a critical role in shaping the news, their influence appears more a result of incremental decision than of grand design.[10] It is the editor who decides what is published, how each article is cut, and where it will appear in the paper.

Journalists, unlike scientists, relinquish virtually all control over the final shape or presentation of their articles. When a journalist hands in a story, the editor first decides whether or not it will be used, and in what form. If it is accepted, it is copy-edited, headlined, and positioned in the paper in the evening—all without consulting the writer, who usually sees the story only when it is published the next day.

Even when relying on the wire services, editors choose and edit stories to fit their judgments about how to maximize reader interest. With the exception of those few who were once science reporters themselves, most editors are trained in the liberal arts and are not very familiar with science. A study of editorial judgment found that editors use different criteria to evaluate science articles from those of either science writers or scientists.[11] Concerned with attracting readers, editors evaluate news stories primarily on the basis of their color and excitement, while science writers' evaluations are based more on accuracy and significance.

It is commonly believed that readers are less interested in analytical reports about science-related events than in how these events will affect them personally. Thus, one editor complains that some science writers on his staff are so preoccupied with scientific approval that they tend to lose sight of who their readers are, whereas he judges articles according to "what my mother wants to read, my kids want to read, my wife wants to read." Even David Perlman, an editor as well as a science writer for the *San Francisco Chronicle,* comments on the pressures of being on "the other side" and having to weigh science articles against news on "politics, murder, arson, fire, theft, corruption, war, revolutions, strikes, hot tubs and the higher California consciousness."[12]

Editorial control over the selection and presentation of articles is often a source of great irritation among journalists. "When they misrepresent the story in a headline," reporters told me, "we give them hell, but we can't do anything about it." The placement of articles is another sore point. One reporter did a story on a large, important research center, only to find it placed on the back page of an edition of his paper that featured the front-page story: "Everything you want to know about pickles." There is, claims another, a sure way to make page one. "Mention in the first paragraph something about a treatment for piles, ulcers or sexual impotence. Every editor has or worries about these conditions."

Another reporter complains that her editor sees science features as the "miracle page" and asks her to avoid writing so many dreary articles. A science writer for a popular magazine says her editor views science as either a "weird filler" or "a class act." He selects material that fits these categories and tries to include them in alternate issues. Editors, many writers claim, tend to change important qualifying words: for example, "may" is often changed to "is" in the interest of style. Or they may inappropriately screen out technical terms: an editor changed a reference to Darvon as an "analgesic" to "tranquilizer," not appreciating that inaccuracy was the cost of using a more familiar word.

It is often the editors who insist on definitive scientific explanations even when there is real uncertainty. One refused to accept a report on Legionnaire's Disease admitting that the scientists did not understand it. He queried the reporter: "What do you mean they don't know? Get them to explain what caused it."

Despite their frequent complaints, science writers say that they have greater autonomy than other journalists. Not only do they generate most of their own stories, as opposed to receiving assignments, but they also enjoy greater freedom to exercise personal judgment in writing these stories: "It's a joy to cover the beat the way I think it ought to be covered"; "I have more freedom as a science writer than I ever did before." And their stories go more lightly edited than those of political reporters. It is because of the specialized nature of their subject that they enjoy such autonomy: "Frankly, if I'm wrong, there's nobody here who will detect that. . . . They just can't challenge me because they don't know. A lot of responsibility is thrown upon myself as a writer."

But science writers maintain this autonomy by exercising self-restraint. Journalists develop a symbiotic relationship with their editors in which self-censorship limits direct editorial control. As one reporter puts it: "Every time I write a story, I always have in mind if it is going to be acceptable." The experienced reporter knows what will be acceptable to an editor and will seldom push his own preferences to the point of confrontation.

Editorial policy, like journalism as a whole, changes over time. During the Nixon administration the adversarial relationship that developed between the executive office and the press resonated throughout the profession of journalism. The press had demonstrated its capacity to exercise political influence: first the Pentagon papers, then the Watergate affair. And in 1994, media coverage of the Whitewater scandal has made it clear that the press is less a means of publicity for those in power than a power in its own right. In response, political administrations are spending more and more money on information services and public relations to enhance their public images. And so too are scientific and technical institutions. Media consultants have emerged as a new profession, dedicated to dampening adversarial tendencies in the press.[13]

Even so, many editors fear that the power of the press will be held against the institution, threatening its First Amendment protection. A 1994 meeting of the American Society of Newspaper Editors dwelled on the "mean spiritedness festering in the journalistic gut today."[14] Editors worried especially about the media efforts to embarrass public figures.

Legal and political decisions are, in fact, eroding some of the freedoms long enjoyed by the press. Journalists lost about 85 percent of 106 major law suits between 1976 and 1983. Though many decisions were reversed on appeal, the threat and cost of litigation has had a chilling effect on investigative reporting. Even in areas not necessarily associated with lawsuits, such as science reporting, the fear of litigation encourages a carefully orchestrated type of journalism characterized by increased use of press releases, adaptation to prevailing political values, and avoidance of sensitive issues. Editorial policy has become, in the words of one editor, "more judicious." Editors no longer "want to turn over a rock just to see what sort of dishonesty is under it." Science writers share these cautious attitudes. They talk about "muzzling themselves," "conforming," "compromising their interests," "self-policing." They avoid covering stories that run counter to prevailing values. As one writer explained: "You have to be practical. What I want to do a strong story on is not always what my editors want a story on. We let a lot of things fall through."[15]

One of the most egregious stories to "fall through" concerned the Department of Energy's (DOE) human experimentation program on the effects of radiation. In 1986, at the height of Cold War sensitivity, the media virtually ignored the massive documentation and intensive efforts of activists and researchers to bring the details of these experiments to public attention.[16] In the early 1980s, for example, a citizens' group called Victims and Veterans Opposed to Technological Experimentation formed to represent those harmed by the studies. And in 1984 and 1985 a network of activists held a series of press conferences revealing their documentation. But the media responded with silence. In 1986, a subcommittee of the House of Representatives chaired by Congressman Edward Markey investigated the DOE program, producing a report called "American Nuclear Guinea Pigs: Three Decades of Radiation on U.S. Citizens." Markey issued a provocative press release stating that American citizens had become "nuclear calibration devices for experimenters run amok." The response? Watered down coverage of the Markey report appeared as back-page news. Then, seven years later, when the DOE chose to disclose the story of its expermiments, the same material appeared in the media as "revelations." No longer constrained by Cold War values and more alert to scientific scandals, journalists wrote dramatic front-page stories "revealing" long-known incidents and questioning the responsibility and morality of the scientists involved.

Audience Assumptions

Editorial constraints reflect perceptions of the public's interests, preferences, and ability to understand complex subjects. Seldom do journalists or their editors receive systematic feedback from readers. Yet, based on their personal observations, they maintain a set of assumptions about their readers and viewers that influences the selection and style of science news.

These assumptions lie behind the reluctance to cover certain politically charged subjects and the preference for stories about discoveries with short-term practical applications. Seeking material that will attract readers and revenues, science magazines in some cases have turned into service magazines, with numerous articles on computers and cameras. The editor of *Science Digest* ruefully explains why. He and other editors had assumed that in a high-tech world people would be interested in reading about science: "In hindsight that was a terrible assumption. For most people, science is some cerebral, lofty intellectual endeavor practiced by ivory tower eggheads in white coats. For most people, it was a required course that ruined your grade-point average and made you feel stupid!"[17]

The effects of perceived audience constraints on the style of science writing are various. David Perlman, for example, likes to use colorful language in his leads: "If you warn readers they are about to read a science story, far too many will skip it as inevitably too complex." While most journalists try to avoid a sensationalist and titillating style, they do tend to magnify events and to overestimate if not sensationalize their significance. Research applications, after all, make better copy than qualifications. "Revolutionary breakthroughs" are more exciting than "recent findings." And controversies are more newsworthy than routine events. Similarly, science becomes more palatable if made humorous or curious: (e.g., "Is Your Nose a Compass?" "How the Lizard Got Away," "Hungry Galaxy That Swallows Its Neighbors"). The film *Jurassic Park* set off a spate of media speculations about the possibilities of cloning dinosaurs that resembled science fiction more than news. As one writer puts it, "We have to mix the pitches—to titillate, educate, inform, and entertain."

The focus on drama, aberration, and controversy in much of the reporting about science and technology reflects the quest of journalists to make their articles more entertaining. Medical stories, for example, are guided by what is popularly known in the profession as Cohn's First Law, named after Victor Cohn, a science writer on the staff of the *Washington Post,* who said, "There are only two kinds of medical reporting: New Hope and No Hope." News about technology and risk, as we have seen, is often presented with all the attributes of fiction, as a story with heroes and villains, conflict and denouement.[18] Complex issues are avoided, for many journalists doubt that their readers will work sufficiently hard to understand them: "It's better to publish pap and get read than to starve"; "Just try explaining the significance of DNA experiments to those more interested in the Giants' standing." And stories that violate common political understanding (such as the DOE experiments during the Cold War) are less newsworthy than those (such as the Chernobyl accident) that reconfirm political stereotypes.

Finally, assumptions about audience preferences influence the format of articles. Television has conditioned people to short bursts of entertainment and glossy information. During the 1970s many observers believed that the press would lose out in the competition with television's instantaneous communication of news and visual impact. The press has survived, but it has become more like television in style. Some newspapers, for example, print shorter articles than formerly, give more attention to visual details, and employ language that is rich in visual imagery. The remarkable popularity of *U.S.A. Today* is attributed to its television format, with very short, profusely illustrated articles. And, conversely, television "magazine" shows have proliferated, presenting snippets of science with colorful visual aids. Newspapers also try to attract readers through dramatic headlines, graphics, and leads. With detailed explanations and qualifications buried deep in the text, the images of science and technology received by casual readers who simply scan the headlines may be quite different from those received by careful readers.

Science writers explain their tendency to oversimplify by arguing that this is necessary to keep their readers' attention: "Those who scream about oversimplification are just not aware of the way people understand things." There are clearly numerous counterexamples: many articles on AIDS research and on the discoveries of genetic markers for disease have been detailed and complex, even though they are directed to a broad lay readership. When an issue holds special interest, many people want technical details. Yet in an age where communication among scientists is specialized and obscure, simplification is an essential if controversial part of making science palatable to the public.

Economic Pressures

Behind the concern about audience interest lies the necessity of profit. Newspapers must maintain circulation, attract advertising and not offend their advertisers, their owners, or their boards. Newspapers are profit-making enterprises; most are owned by large newspaper chains, and sometimes these are bought out by multimedia companies and foreign firms.[19] A study of the boards of directors of the 25 largest newspapers found that most of their members have ties to the corporations and universities that their papers cover regularly.[20] Moreover, newspapers must operate according to the commercial realities imposed by their dependence on advertising. A typical newspaper derives about 80 percent of its income from advertising, which consumes about 65 percent of newspaper space.[21]

The fate of science magazines and the science sections of newspapers rests more on computer, camera, and video ads than on subscriptions. It was a decline in advertising revenues, not a loss of reader interest, that led to the demise of several science magazines (*Science Digest and Science 86*) in 1986 and to the decline in newspaper science sections in the 1990s. To sustain the necessary advertising level, the press is constantly struggling to attract readers—appropriate readers from a social class that generates advertising income.[22]

Commercial constraints on what is published are seldom directly imposed. An editor, William Allen White, once argued, "The publisher is not bought like a chattel. . . . But he takes the color of his social environment."[23] The editors and publishers of newspapers, especially in small towns, are often friends of corporate executives and may be constantly exposed to social pressure. One editor who joined a local country club described how the people he met there always wanted "to get something into a newspaper or to keep something out." For the reporter in this sensitive economic and social context, caution encourages routine, noncritical interpretations: "We are trained subtly not to cover the corporate sector and other private institutions the way we cover City Hall."[24]

The pressures on the policies of local newspapers are apparent if one compares the coverage of similar events in different areas with different economic stakes. Recall the cautious accounts of the ozone risks in Delaware, a state economically dependent on Dupont. In another case, when 175 children in a northern Idaho town were found to have dangerous concentrations of lead in their blood, the region's newspapers shied away from the story. Pressure from the local lead and zinc smelter contributed to the news blackout. Moreover, scientists in the area were mostly employed by the industry and would not help reporters unravel the issue. Those stories that appeared had headlines minimizing the problem: "Doctors Said Lead Scare Out of Proportion," "Lead Poisoning Fears Largely Unwarranted," "Bunker Hill [smelter] Warns Regulation Could Cause Shutdown."[25] Media analysts have documented the influence of community power relationships in shaping the media coverage of environmental risks.[26] They sometimes refer to the "Afghanistan syndrome"—the tendency to cover problems occurring in distant places or to think of problems as most serious in regions that are "up the road a piece."[27]

Some science reporters view these trends with cynicism but either restrain themselves or are restrained by their editors from bucking local pressures. Some typical comments: "When I got my first job in the newspaper, I saw all the Woodward and Bernstein movies and I read all the books on how moralistic and ethical newspapers were. The one thing I've learned is that newspapers . . . can be just as ruthless as any other company." "I cannot even call a company a company without comment from the company, because we're so afraid of losing ads." "If you are in a one-industry town, to say that has no influence is naïve."

External pressures are often indirect, but many science writers feel constrained by corporate pressure at the national as well as the local level: "I pay hell for doing consumer-related health issues." "Companies put pressure on my paper all the time. If I don't comply, I'd be punished." Economic pressures are even more telling on television, which obtains its revenue from selling air time based on viewer ratings. Programming reflects efforts by the major networks to gain the widest possible audience and to avoid displeasing the corporate interests that provide advertising revenues. Thus the form and content of science information on television is conditioned by what producer Jeffrey Kirsch calls "the marriage of the sales mentality to the electronic image. . . . Producers have learned what formats, production, techniques, symbols, and personalities are most likely to succeed."[28]

Constraints of Complexity

The constraints of newswork, editorial supervision, audience assumptions, and economic pressures are faced by reporters in virtually every field, although the nuances may differ. Peculiar to science writing, however, is an additional constraint—that of having to assimilate and simplify vast amounts of sometimes extremely complex material. About 50,000 journals and about one million scientific papers are published every year, and rapid obsolescence of papers has increased with the growing quantity of information. In certain active fields, some 70 percent of the citations in scientific papers refer to research only during the preceding five years.[29] This so-called information explosion has many consequences for journalists, who, even if scientifically trained, cannot possibly keep up with the latest details of all specialties.

Journalists covering a science issue may not know how to recognize what is important and may therefore miss a newsworthy story. They try to prevent this by keeping an eye on the more experienced science writers, but few journalists (or their readers) can judge if numbers are meaningful and sampling techniques or research methods appropriate. Furthermore, explanations in science often defy common sense. As a journalist puts it, "The news part is not hard to get right. . . . It is the mechanism by which something works that is hard to grasp and to simplify. How does the birth control pill work? How do you measure the rate at which the universe is expanding?" Yet complexity requires such explanations; readers will need background material if they are to understand the significance of scientific discoveries or events.

The complexity of technical information compounds the usual problems of maintaining accuracy in reporting—an extremely important value within the system of the press. Good journalists check their information with care. They sometimes help each other with explanations of difficult concepts or controversial issues, using each other as experts at press conferences. "We rely on each other's background and knowledge." "We keep each other apprised of what's going on and even sometimes help each other with a black sheet [a copy]." Yet scientists repeatedly complain about the inaccuracy of science reporting: "Whenever I read about something in my field, it is usually wrong."

Studies assessing the extent and possible causes of error have produced mixed results.[30] They suggest that 40 to 50 percent of scientists complain about inaccuracy. When pressed, however, they usually identify the problem as one of omission of relevant information and lack of qualifying statements rather than of error per se.[31]

Often errors derive less from inaccurate reproduction of details than from distortions that occur in translating complex technical terms into lay English. But they may also result from the writer's limited sense of the technical intricacies of an issue. For example, in covering the research linking saccharin to bladder cancer, many journalists failed to understand a standard scientific procedure based on extrapolations from large-dose experiments, and therefore misinterpreted the research.

The technical uncertainties characteristic of so many newsworthy issues in science compound this problem of complexity.[32] To report on health risks journalists must sort out uncertainties concerning the extent and significance of problems, their causes, the potential damage resulting from exposure, and the appropriate methods of remedial action—all of which may be subject to conflicting scientific interpretation. In many instances information simply is not available. Risk from toxic chemicals is a case in point. Relatively few of the 63,000 chemicals in commercial use have been tested for their chronic health effect.[33] Compounding technical uncertainty is the difficulty of drawing precise causal associations between exposure to toxic substances and long-term health consequences. Neither epidemiological nor laboratory studies can identify all the substances that may cause cancer, neurological problems, or genetic defects.

The problem of associating cause and effect contributes to disputes over the impact of dioxin exposure on birth defects and spontaneous abortions. It confounds analysis of the health effects of food additives, the influence of ozone depletion on plant growth and human health, and the relationship of diet to cancer.[34] In these and other cases scientific uncertainty allows a wide range of interpretation of risk. With few standards to guide their analysis, the specialists assessing risk arrive at different conclusions about the relative danger of particular hazards.

For journalists trying to write about technical problems and to assess their significance, conflict among scientists presents a dilemma. Their difficulties are compounded by technical jargon and the excess information that scientists sometimes provide as a way to create the illusion of certainty and competence in confusing situations. The quality of reporting on risks varies, reflecting the journalists' technical sophistication and ability to sort out and interpret the available information. But, as we have seen, reporters most often simply try to present a balanced account by citing all sides of risk disputes without evaluating the quality of competing claims.

The complexity and uncertainties of scientific subject matter reinforce the tendency of journalists to rely on press releases, press conferences, and other prepackaged sources of information. Journalists attempt to compensate for the inadequacies of their own judgment by cultivating scientists who will give them a "scoop." But as historian Anthony Smith suggests, "One man's scoop is frequently another's carefully camouflaged handout."[35] Difficulty in coming to grips with complex technical material reduces the likelihood of skeptical, probing investigation. But even more important, complexity converges with the constraints of journalism to increase the science writer's vulnerability to those sources who try to shape the news.

Vulnerability to Sources

In May 1981, 175 newspapers around the country published a UPI wire service story about a scientist, Dr. Gregor, from a research institute called Metamorphosis, who had discovered a wondrous, all-purpose cure made from cockroach juice. The "scientist" had circulated press releases and announced his "findings" at a press conference. That reporters and editors believed his hoax could have reflected their ignorance of Kafka, or, perhaps, their respect for the invulnerability of the cockroach. But it is also a telling illustration of the vulnerability of many journalists when confronted with scientific and technical sources of information.[36]

Science journalists receive an enormous amount of mail. One writer told me that he scans 58 journals a month, receives about 250 press releases and 40 letters a week, and answers about ten phone calls a day from scientists or their public relations officers who believe they have science news. He also attends about twenty scientific meetings a year and two press conferences a week. Another reporter said he finds several feet of press releases piled high on his desk every day.

Press releases are an American invention.[37] They first appeared in 1907, but only in recent years have they commonly been used as a means of shaping the news. Inundated with press releases, reporters are often cynical about them. Joseph Alsop once said, "All government handouts lie; some lie more than others."[38] A science writer for *Time* feels that they are "useless," "fit for nothing more than the garbage pail."[39] Another suggests that "every good reporter needs a good shit detector, and this holds true especially for science writers."[40] Still another finds press releases sad: "Sad for the eager publicists and their employers who undoubtedly labor hard to create the discarded press releases. Sad for the postal workers who must post this bundle of mail each day. Sad for those of us who must sift through the numerous letters. Above all, sad for the trees."[41]

Nevertheless, when time is short and information complex, reporters rely on press releases, often adopting their language as well as their content. In a study of sources of environmental information used by the media, David Sachsman found that during an eight-week period eleven environmental reporters received 1347 press releases. They relied heavily on these sources: over 50 percent of the published stories could be traced to these public relations efforts, and in many cases the reporters simply rewrote the release. Writing science news, Sachsman concluded, often involves simply "opening the mail."[42]

Science reporters also rely on information from press conferences at professional society meetings. These, too, they regard with some cynicism as "preselected and controlled news," or as "a source of free meals, free booze." They note that the American Cancer Society meeting is always timed to coincide with fund-raising campaigns. They complain of the constant "hype" about new cancer cures. "It won't be in clinical use for ten years and they give you the impression that it will be."

Although they feel manipulated, science writers admit their dependence on such preselected news. Most journalists covering a scientific meeting will report on the same stories. As one reporter put it, "In the presence of your fellow creatures . . . somehow it magically works out that everybody tends to agree on what is important." Inexperienced writers are most likely to rely on press conferences and to passively accept the material provided. More experienced writers use press conferences to meet people and to gain ideas, then actively pursue stories and check them out through their personal contacts.

Who are these contacts? How do reporters find reliable sources and assess where they stand in the world of knowledge?

The technical nature of science encourages reliance on official sources of information—predictable sources who know how to package information for the press. Leon Sigal, documenting the sources of information used by the *New York Times* and the *Washington Post,* found that 46.5 percent were officials in government agencies, 4.1 percent were from state and local agencies, and 14 percent from nongovernmental groups.[43] A content analysis of the press coverage of the environmental controversy over PCBs in the Hudson River found that 43 percent of the sources cited were government bureaucrats. The *New York Times,* however, was more likely than local papers to cite scientists rather than public officials, because the specialized reporters on the *Times* had many contacts in the scientific community and greater confidence in dealing with the technical information they provided.[44]

Experienced science writers know many scientists personally and have a "stable" of trusted experts, "inside dopesters" on whom to rely. But their sources do not necessarily represent the spectrum of opinion. A *New York Times* reporter told me in a somewhat embarrassed tone, "I prefer to interview those people I agree with the least. When I interview environmentalists, they offer insufficient hard data for me to publish their views. They may be right, but they have a conspiratorial attitude: there's a snake under every rock. They feel they are morally right and it's hard to get them to be specific. Industrial scientists, on the other hand, have data up to their eyeballs. They're smooth, they know what the media needs."

Less sophisticated reporters will often turn for expertise to those scientists who happen to be most accessible, often those teaching in a local university. Usually the reporter cannot assess the experience of these scientists, their knowledge of the subject, or their reliability as sources. Faced with technical claims that are difficult to check out, and socialized to regard scientists as a reliable and objective source of information, these journalists are inclined to believe what they are told.

Experienced or not, science journalists are constrained by their concern about future access to scientific sources. A *New York Times* journalist feels constrained less by editors than by the community she covers. "I've come from the ranks of science writers recently to join the political spectrum covering HEW for the *Times* in Washington, and I feel a lot freer covering politics than I ever did covering science. . . . It is very difficult [for science writers] not to be on the team. I am allowed to say things about the President, using my basic instincts as a journalist. That you just wouldn't think of doing in science."[45]

The vulnerability of science journalists converges with the economic and social constraints of newswork to give an unusual degree of power to those best organized to provide technical information in a manageable and efficiently packaged form. Scientists and their institutions, increasingly motivated to enhance their influence over science and technology news, have become more sophisticated about how to do so. They exercise their influence primarily through two effective strategies: expanded public relations efforts, and increased controls over the dissemination of information to the press.

8

THE PUBLIC RELATIONS OF SCIENCE

In his 1993 presidential address to the American Association for the Advancement of Science (AAAS), F. Sherwood Rowland attributed the most serious problems of scientific progress to "faulty communication." He called for more science in the media because the public that supports science "seems less than fully appreciative of what we see as our tremendous successes." He urged his colleagues to sell the importance of science through better communication.[1]

Scientists ventriloquate through the media to those who control their funds. In 1985, Nobel Prize winner Kenneth Wilson convinced the National Science Foundation to support the supercomputer program. He attributed his success to a prestigious scientist's warning that without a nationally funded program the United States would lose its lead in supercomputer technology. "The most amazing thing to me," said Wilson, "was how the media picked it up and [how] a little insignificant group of words could change everything." This group of words had worked by establishing a politically powerful image. Wilson thus learned a basic principle in public relations:

"The substance of it all [supercomputer research] is too compli-cated to get across—it's the image that is important. The image of this computer program as the key to our technological leadership is what drives the interplay between people like ourselves and the media and forces a reaction from Congressmen."[2]

Traditionally working in a context where success is measured by the judgment of peers, scientists have long assumed that a record of accomplishment is sufficient to maintain research sup-port.[3] Thus, information, the scientist's "stock-in-trade," has been directed primarily toward professional colleagues. Most scientists have not been interested in public visibility; on the contrary, they have feared it could result in external controls on their work. But attitudes in the scientific community have changed. Depen-dent on corporate support of research or direct congressional appropriations, many scientists now believe that scholarly com-munication is no longer sufficient to maintain their enterprise. They see gaining national visibility through the mass media as crucial to securing the financial support required to run major research facilities and to assuring favorable public policies toward science and technology.

This is not an entirely new perception; public relations in science goes back to the nineteenth-century, when the new pro-fessional societies encouraged popularization of science as a way to build "public appreciation of the benefit that science provides to society."[4] As Rowland's address suggests, the motivation is the same, but the scale of public relations efforts has vastly increased as the media are perceived as a major vehicle to enhance the public support of science.

Promoting Scientific Institutions

In 1847 Joseph Henry, the physicist and first secretary of the Smithsonian Institution, spoke of the need for greater public representation of science: "In carrying out the spirit of the plan, namely that of perfecting men in general by the operation of the Institution, it is evident that the principal means of diffusing knowledge must be the press."[5] In the nineteenth century, scientists such as Louis Agassiz, T. H. Huxley, and John Tyndall actively publicized their own work by giving popular lectures and writing directly for the press. With the increase of private philanthropy and industrial support at the turn of the century, however, scientists returned to their laboratories; popularization declined, and indeed began to appear unseemly as a professional activity.[6]

After the First World War an expanding scientific enterprise needed greater public funding, and once again scientists sought ways to enhance their public image. Professional associations began to organize news bureaus and public relations departments. In 1919 the American Chemical Society became the first scientific association to organize a news service. It hired a professional science writer to translate technical reports for the public and to write descriptions of scientific research for the press. As new associations formed in the 1930s, they followed the chemists' lead.

Recognizing the importance of the press in creating a favorable public image, leading scientific organizations such as the National Academy of Sciences and the AAAS were willing and eager to collaborate with Edwin Scripps when he founded the Science Service in 1930 (see Chapter 6). That decade also brought systematic public relations efforts from the medical profession. In 1937 the American Medical Association tried to establish rapport with science writers in order to counter the growing appeal of quack medicine. It created a press relations office run by a journalist, Lawrence Salton. Also during the 1930s, the American Society for the Control of Cancer used the press to implement a nationwide campaign to encourage early detection of cancer.

Public relations activities by scientists increased again in the years immediately following World War II. In 1952 the American Cancer Society (ACS) began the first of a series of laboratory and hospital tours for journalists, a veritable oncology marathon, to interest them in the goals of the ACS. The tour program, which continued for seven years, included visits to the universities, hospitals, and medical centers participating in ACS-sponsored research. One group of reporters visited centers in 21 cities in twenty days. The schedule prompted a poetic response, "The Saga of the Sarcoma Special":

> We heard of work on mouse and frog and newt and toad
> and spider,
> And bee and fly and bird and chick and hen and Arthur
> Snider.
> We heard of cortisone and glands, ACTH, anemia,
> Pituitaries, ovaries, adrenals and leukemia,
> Metastases, removals, series blocks and ugly rumors,
> The breast, the pancreas, the rays that never cure the tumors,
> The palliative surgery which might destroy the bone,
> The seven danger signals which all seemed to be our own,
> And when we'd heard it all from radiation to mitosis,
> We stood at last as experts in one field:
>
> Tuberculosis[7]

As federal research funding expanded during the postwar period, interest in popularization once again waned. But the 1957 Soviet launching of *Sputnik* brought renewed concern about the gulf between science and the public, with its implications for America's leading role in international affairs. Harvard chemist George Kistiakowsky expressed what many scientists felt at that time: the need for "the skillful interpreter who can translate scientific results and findings into language that the average reader can understand and appreciate."[8] Scientists again discovered that, as a public relations officer put it, "the surest way to capture a share of the funds was to do good research and, almost as important, to talk about it."[9]

Scientists in the ensuing years supported the popularization of science out of ideological and cultural as well as economic concern. In 1960 Jean Rostand, a French biologist and popular science writer, expressed a prevailing view: "The true and specific function of popularization is purely and simply to introduce the greatest number of people into the sovereign dignity of knowledge, to ensure that the great mass of people should receive something of that which is the glory of the human mind . . . to struggle against mental starvation and the resulting underdevelopment by providing every individual with a minimum ration of spiritual calories."[10]

Others, such as Jacob Bronowski, the British biologist and popularizer, talked of the need for a "democracy of the intellect." "We must not perish by the distance between people and power that destroyed Nineveh, and Alexandria, and Rome," Bronowski wrote. "The distance can be closed only if knowledge sits in the homes and heads of people and not up in the isolated seats of power."[11]

As the scientific and technical enterprise grew in complexity and importance and the demands for research funds began to outstrip the supply, scientists increasingly emphasized the pragmatic goals of public communication. In 1971 a conference of biologists and health scientists, for example, concluded that the public must be given sufficient background material to understand the practical payoffs of seemingly abstruse basic research so that it would be more willing to provide research funds.[12]

Since the 1960s, professional societies, academic institutions, and research organizations have all increased their public relations activities in order to enhance institutional prestige, encourage public support of research, and influence public policy with respect to science and technology. For example, the American Institute of Physics (AIP), founded in 1935, expanded its publicity programs in the 1960s, running seminars for journalists and news conferences to summarize newsworthy developments in physics.

The AIP routinely issues press releases and provides instruction to physicists on how to deal with reporters. The National Academy of Sciences, which in the past primarily explained and interpreted technical reports for interested journalists, changed the style of its media relations in the 1980s, assuming a far more active role by initiating press releases and seeking maximum coverage of their reports. Its staff regards the press as a way to shape public attitudes and to influence congressional decisions on the funding of science.

Government agencies involved in costly technological or scientific developments play the same game. NASA's highly sophisticated public relations apparatus, intended to win popular support for its costly program, was so successful it completely diverted the press from issues of safety and administrative mismanagement. It took the *Challenger* accident to call attention to such issues and to the role of NASA's public relations: "Some agencies have a public affairs office; NASA is a public affairs office that has an agency."[13]

Scientific journals also see the advantage of media publicity. Advance copies of the *New England Journal of Medicine and Science* are sent by first-class mail to journalists, who must respect the mandated release date before writing stories on the articles. These competitive journals want to maintain their image as the key sources of scientific information for the public, and they skillfully use the press to this end.

Individual scientists try to attract press attention for a variety of reasons—to influence public views, to attract funds, or to establish their competitive position in "hot" fields of research. In 1977 DNA researchers responded to the dispute over the safety of recombinant DNA research by initiating a remarkable media campaign to show that genetic engineering research was safe, its critics irresponsible, and regulation unnecessary. Geneticists today, seeking to maintain support for costly research, have become skilled in rhetorical strategies designed to attract the media.

They describe the genome as a "bible," a "medical crystal ball," a "blueprint of life." They promise that the Human Genome Project will "unlock the secrets of life," allowing the prediction and control of disease.

Computer scientists also make extravagant claims to attract public support. Edward Feigenbaum, an artificial intelligence expert, wrote that with the fifth generation of computers "revolution, transformation, and salvation are all to be carried out."[14] Physicist Kenneth Wilson called the new computer development "a second renaissance."[15]

Individual scientists sometimes promote their own work, but they usually rely on their institutions to disseminate information to the press. Most major research universities employ public relations professionals or outside media consultants to publicize the work of their science faculty and thereby to enhance the image of their institution. Good public relations is important to these institutions, which must attract good students and staff, obtain money for research, and maintain public legitimacy. Their public relations professionals may be experienced science writers themselves, but unlike reporters they work for those they write about. Their job is to insure that their institution's research is covered prominently, accurately, and favorably in the press.

Public relations in medical institutions play to a receptive press eager to publish stories on the most advanced transplant experiment, the latest gene therapy, or the newest research on reproductive technology.[16] Some research—the impregnation of postmenopausal women, the cloning of embryos, or the harvesting of eggs from aborted fetuses—has attracted media attention more for the titillation than the significance of the science. But the valued media coverage has also attracted the unwanted attention of antiabortionists.

Dramatic personal stories also appeal to the media. In 1982, ten-month-old Jamie Fiske was admitted to the University of Minnesota Hospital for a liver transplant, but no donor was in sight. Her parents went to the press with a dramatic plea. The case had all the elements of a good story: a personal tragedy, a family in despair, a father fighting bureaucratic obstacles to save his child, and the possibility of salvation through the wonders of medical science. The press responded, and news articles appeared throughout the country. Their effect? A liver for Jamie Fiske, a letter from the President, and money for both the family and research. Benefits spilled over to other transplant units, which enjoyed an increase in research funds as well as donated livers, including some offers from living people who, unfamiliar with anatomy, offered one of their own.[17] The pattern set by Jamie Fiske has continued and the media have become a vehicle for patients seeking donors or funds for medical care. But media solicitations have also backfired; in one case a TV station embarked on a crusade to raise money for a child with leukemia, and $150,000 was collected before a reporter discovered it was a scam.[18]

As scientific and medical procedures became increasingly costly during the 1980s, technical institutions expanded their public relations efforts. The most dramatic example was the extraordinary promotion of the artificial heart. In 1982 the Utah Medical Center employed a team of public information officers to attract media coverage of the first human implant of an artificial heart. Hundreds of journalists responded. When the artificial heart team moved from the Utah Medical Center to Humana Hospital in Louisville, Kentucky, this for-profit medical care system provided even greater resources, hiring a public relations firm and creating a center staffed by seven professionals to handle press coverage. The staff passed around briefing books with exhaustive information on the personal as well as the technical aspects of the case. They provided telephones, places to work, and even food. It was, in the words of one reporter, "like covering a football game where they hand out statistics at the end of every quarter."[19]

The remarkable media exposure served the hospital well, enhancing its reputation and helping to fill beds. But after the initial burst of publicity, hospital physicians began to talk of the "media onslaught," describing it as "intrusive," "invasive," and an "ordeal." As the patient's condition worsened, Humana's approach to the press, according to a reporter on the *Louisville Courier Journal,* turned from "NASA style" promotional publicity to "Soviet style" silence, which in effect put the experiment under wraps.[20]

Artificial heart experiments have ended and heart transplantation procedures are now routine and low key. But the pattern of promotion continues in other areas, encouraged by the growing commercial interest in academic science and the dependence of researchers in fields such as genetics and biotechnology on the support of industries interested in their products. The zeal of science publicists can, however, generate misleading reports—especially about medical research. Research on possible therapies for a devastating disease is always newsworthy and often publicized far too soon. For example, in 1984 public relations officers at the Dartmouth Medical Center held a press conference at which investigators announced the results of a preliminary feasibility trial of a therapy for Alzheimer's disease. The research had been tried on only four patients, as explained in a published paper. But the press release failed to mention the study's limitations, and the very decision to hold a press conference turned the research into a media event. Not surprisingly, the press headlined the research as a "breakthrough," a "successful treatment," and a "possible cure," raising the hopes and expectations of readers and sick people desperate for information. Twenty-six hundred people called the Dartmouth center to inquire about a cure.[21]

Scientists in the Human Genome Project are enthusiastically publicizing their discoveries of genes linked to common diseases such as breast and colon cancer, for this helps to justify support of their costly research program. But their promotion has also raised public hopes and expectations about therapies that are not yet in sight. Seldom are readers warned that the ability to test for a genetic disease may have little to do with intervention.

Scientific experiments in other areas of potential public or commercial interest have also been promoted in misleading ways. Perhaps the most infamous example in recent years was the work on cold fusion at the University of Utah in 1989. There, two scientists, B. Stanley Pons, Dean of Chemistry at the university, and Martin Fleischmann, of the University of Southampton (England), reported achieving hydrogen fusion in an electrolytic cell, obtaining a substantial yield of energy. The university announced the research at a press conference. "We couldn't afford to wait the six to eight weeks for publication," said the Vice President for Research. Thus the results appeared, not in a peer-reviewed journal, but in media announcements of "a tremendous scientific advance" that would assure "the almost limitless clean power of nuclear fusion."[22] The university withheld technical details due to "patent problems." The media coverage had its intended effect, attracting venture capital and a commitment of $5 million from the state legislature. The scientists enjoyed immediate, though short-term, fame. Several months later their experiments were discredited by scientists unable to replicate them. Embarassed and confused, reporters withdrew from the technical story, focusing instead on personal issues such as the scientists' troubled relationship with the university. Subsequent reporting marginalized Pons and Fleishmann and turned them into villains.[23]

A common pattern marks these and other cases. Journalists, looking for a dramatic story and pressed for time, are inclined to believe their scientific sources and to rely on public relations professionals. But even when reporters suspect that publicity-seeking lies behind dubious scientific claims, they may feel compelled to publish them. For example, all during the early fall of 1985 articles on the AIDS virus were appearing daily in the press, relating stories of awful illness and frustrating efforts by scientists in the

United States and France to find cures. In November of that year a group of French scientists announced a "successful" trial of a drug that they claimed had reversed the course of the disease in several patients. Experienced science reporters were dubious. They knew that a trial on only a few patients was unlikely to be significant and that the time elapsed since the drug had been administered was too brief to mean very much. Moreover, in light of the intense international competition among AIDS researchers and the commercial potential of a new drug, they suspected the motivation of the announcement. Some reporters preferred to ignore the story, but what were they to do? Given the public appetite for AIDS news, the preoccupation with the disease, and the competition in the news business, scientific claims about this important subject could not be ignored. Yet there was no way to check out the research. Reporters felt they had no choice but to report the information as provided by the researchers—and few had the confidence to report it with the skepticism it deserved.

In another case, in 1993, scientists went to the press too early with reassuring findings that the risks of cancer among women with a family history of the disease were less than previously assumed. Journalists, aware of the public interest in such find-ings, enthusiastically reported the story, but only three months later they were told there had been an error in the study.[24]

Scientific promotion is a continuing source of concern to many science journalists. As reporter Robert C. Cowen of the *Christian Science Monitor* put it: "We can deal with the Union of Concerned Scientists on one side of an issue and, knowing that bias, can deal with its other side. But when the bastions of profes-sional purity and objectivity begin to worry about budget, jobs, and image to the point of reducing themselves to song and dance, where can science writers turn for objective background on scientific developments?"[25]

Some editors feel that their newspapers are used as pawns for grantsmanship. "When government money was available easily," an editor claims, "you couldn't get a story out of a molecular biologist. Today, I get copies of grant applications in the mail with this thing, 'single cure for blank,' or whatever the hell it might be, circled in red, saying 'we need all the help we can get, fellers.'"

Scientists in Industrial Public Relations

Just as academic institutions sell the importance of their science to attract a favorable press, so corporations use the prestige of science to enhance their own goals. There is, of course, a long tradition of using scientific images in advertising to enhance public confidence in products. But firms now also use scientists themselves in their public relations efforts.[26]

Industrial public relations developed at the turn of the century, first as an adjunct to advertising and later as a response to crises that could damage corporate reputations. Du Pont, for example, formed its public relations department in 1934 after a Senate investigation of the gunpowder industry created an image of the company as "a merchant of death." Following the saccharin dispute, the Calorie Control Council spent about $3 million a year on public relations for the artificial sweetener industry. And since the Surgeon General's report on the harmful effects of smoking, the tobacco industry has spent enormous resources to counter its negative image in the press.

Damage control remains a major goal for public relations, and scientists are engaged to enhance corporate credibility and legitimize company claims. Richard Tucker, a scientist and the president of Mobil Diversified Businesses, suggests why: "In an atmosphere like that of Times Beach or Love Canal or Three Mile Island, what has been generally missing is the voice of the calm, responsible scientist. To combat unreasonable fear, scientists must communicate more with the public and the media. . . . Public fear of chemistry is unlikely to be abated unless we try."[27]

One means of engaging scientists appears in a kind of advertising placed on the editorial pages of major newspapers. Called "advertorials," these columns resemble news items or editorials more than ads in format, and they deal with contemporary issues such as government regulation or environmental health and safety. The chemical industry, for example, advertises its competence and public concern by printing photographs of scientists or engineers (sometimes accompanied by their children). Some of the captions are as follows: "My job is managing chemical industry wastes. What I do helps make the environment safer today and for generations to come"; "As a chemical industry engineer, I work hard to keep my community's air clean. After all, my grandchildren breathe it too"; "We will engineer out risks, impose detection techniques, expand studies"; "We have a technical staff of 10,000 specialists whose job it is to protect the environment."

Some companies employ scientists to influence the coverage of controversial technologies or products. During the disputes over the safety of nuclear power plants in the 1960s and 1970s, the nuclear industry developed an elaborate public relations apparatus engaging scientists to "enlighten" the press.[28] The idea was to "use the right medium to communicate the right message to the right target audience." Scientists were the medium: "The public has faith in science, believes scientists and would listen." The campaign should "put articulate scientists out front."[29] Westinghouse sent engineers and scientists around the country to lecture and make themselves available for interviews with reporters. Public relations firms that specialized in running political campaigns trained the scientists for public debate and taught them how to approach the media.

The chemical industry has used similar strategies to counter its negative image as a risky business. Feeling "under siege" from a medium that intends "to lay bare the supposed transgressions of the powerful against the powerless," companies use scientists to provide "the facts" that might rebut "erroneous stories."[30] Public relations officers argue that while the media disseminates material to create irrational fear, hysteria, "cancerphobia," and even "chemophobia," *they* use the press to teach the public that chemicals are natural, benign, and essential to life: "We must get across to the public the value of chemicals in our lives. . . . Scientific organizations and chemical companies alike must renew their efforts to find audiences to hear their story."[31] Dow Chemical established a "visible scientist program," sending scientists from the company on media tours. Dow scientists ("credible scientists, not corporate spokesmen," according to company brochures) talk to civic groups and reporters in areas where chemical issues are of special public concern (for example, near major chemical plants or toxic waste sites).[32]

Observing that scientists have greater credibility than corporate clients, public relations firms have advised corporations to develop "parachute teams" or "truth squads" of scientists, ready to move into risk situations in order to defuse the opposition by presenting technical "facts."[33] A New York City public relations firm, Hill and Knowlton, ran a "visible scientist" program for corporate clients, arranging meetings between corporate scientists and the "right" editors. This firm orchestrated the Calorie Control Council's public relations campaign in opposition to the proposed saccharin ban. It was engaged by Metropolitan Edison to increase its press credibility after the Three Mile Island accident and by Ayerst Laboratories to offset the negative publicity about risks of estrogen replacement therapy.

Scientific conferences on risk and the media are another vehicle for public relations. Often jointly supported by corporations and universities, their ostensible goal, as stated in the brochure of a conference on the reporting of technical information on toxic substances, is "to provide a forum for open discussion of the various problems encountered by the media in obtaining and

137

communicating accurate, well-balanced information on toxic substances to the general public." Industrial scientists and public relations officers, however, are the dominant voices at such meetings; seldom are neighborhood activists, union representatives, or other critics to be heard. Systematically focusing on the "problems" of the press in communicating technical information, these discussions are often simply a thinly veiled effort to create a science-based consensus that is compatible with the industrial agenda. By organizing the meetings, the sponsoring universities lend their credibility to promotional efforts.

In 1991 the Chlorine Institute helped to support one such conference on dioxin at the well-known Banbury Center at Cold Spring Harbor, New York. Given the prestige of this center, the meeting attracted scientists who would not normally have attended an industry-supported meeting on this controversial topic. After the meeting, the Institute's public relations office disseminated its results to the media. The press packet included a "consensus summary," not reviewed by participants, asserting that the meeting "reinforced the notion that dioxin is much less toxic to humans than originally believed."[34]

The use of science in public relations is not limited to industrial interests. Environmental and science advocacy groups also engage scientists to enhance their credibility and to legitimate their point of view. Members of the Union of Concerned Scientists tried to keep in touch regularly with reporters during the debate over nuclear power in the 1970s. Scientists such as John Gofman and Helen Caldicott used their credentials to win credibility for the antinuclear movement. Caldicott created a style attractive to the press by publicizing herself as both "scientist and mother." Antinuclear scientists worked with folk singers and rock artists to attract the media and drew Nobel laureates into the debate to legitimate their position.

In the 1990s, the animal rights movement has adopted similar strategies in opposing the use of animals in research. Entertainment stars appear at public events to attract attention to their cause, and sympathetic scientists write tracts for the movement to legitimize its demands for alternative research methods. Animal rights activists use their scientists to attract the media, to create public attitudes sympathetic to their point of view, and to neutralize the arguments of those other scientists who stress the importance of animal experimentation.

Science is also a marketing resource. Corporations often try to sell their products directly through the press by publicizing new therapies as newsworthy scientific discoveries or significant medical advances. The science-based publicity about estrogen replacement therapy described in Chapter 3 has significantly expanded. By 1992 Wyeth-Ayerst Laboratories was spending over $9 million to advertise their product in women's magazines that, appealing to their aging readership, often feature articles and news reports on the effect of ERT on the symptoms of growing old.

This is part of a history of selling products through science journalism. In 1982, Lilly's arthritis drug, Oraflex, had also been marketed through science-based public relations directed at the press. The firm's public relations office sent out 6500 press kits, promoting this new drug by making scientific claims of its effectiveness in relieving arthritis. Some experienced science reporters refused to cover the story, suspecting that Lilly's assertions were exaggerated. However, Oroflex was covered as science news by 150 newspapers and television stations, and prescriptions increased from 2000 to 55,000 a week. When a report showed its harmful side effects, the Food and Drug Administration (FDA) intervened, and after only twelve weeks, Oraflex was withdrawn from the market.

Announcing "medical breakthroughs" is, not surprisingly, a way for companies to increase the sale of commercial projects. Retin-A (claimed a *Journal of the American Medical Association* editorial called "At Last a Medical Treatment for Skin Aging") is a scientifically proven way to reduce wrinkles. The story was widely covered in the media; "A medical milestone about as eagerly sought as the fountain of youth appears to have been reached," said the *Los Angeles Times*. Johnson and Johnson (the maker's) stock increased by 8 points in two days.[35]

Individual scientists have also used their credentials to market products. Several medical researchers promoted the use of cortisol antagonists in the treatment of anorexia directly through the media. Rather than submitting their work (based on a limited study of only 33 patients) to a scientific journal, they announced the findings at a press conference and on a television talk show. "One can't afford to take the time it takes through the medical journals," said one of the doctors. The rush, according to an article in *Forbes,* was related to their effort to market a proprietary line of nutritional products and to expand a private anorexia clinic.[36]

In June 1984 David McCarron, a scientist with the Oregon Hypertension Program, published a paper in *Science* on research indicating that low levels of calcium in the diet are sometimes associated with high blood pressure (hypertension) and that calcium supplementation beneficially reduced blood pressure. McCarron's research, partly supported by the National Dairy Council (a trade research group for the dairy industry), was controversial, and a subject of debate in the scientific literature. Other studies of the relationship between diet and hypertension pointed to the side effects of calcium supplements and argued that the data were too limited to warrant dietary recommendations. Yet at the behest of the Dairy Council, McCarron hired two public relations firms to promote the *Science* article in the press.[37]

Public relations professionals see themselves as "an important arm of the media," a means to save editors hours of work tracking down the news.[38] Their professional society defines its principles as follows: "In serving the interests of clients and employers, we dedicate ourselves to the goals of better communication, understanding, and cooperation among the diverse individual groups and institutions of society."[39]

In the area of science, public relations officers do contribute in important ways to informing the public about products, ideas, and services. They can be a useful source of information for journalists. Presenting complex material in a manageable form, they serve as liaisons between scientists and journalists, easing the job of reporting science. They are, in effect, "marriage brokers," bringing together scientists and science writers, teaching each group how to approach the other.[40] However, what journalists gain in efficiency they may lose in reliability. Public relations officers must make their clients look good. They know how the press works and use this knowledge to promote the interests of the institutions that employ them. Thus, many reporters see public relations less as a source of information than as "a means to promote, protect, and enhance the image of an institution, company, or product."[41]

From the earliest days of public relations, journalists have regarded such efforts as a means to subordinate journalism to private interests. In 1919 Frank Cobb of the *New York World* complained that direct channels of news were increasingly closed as information was filtered through publicity agents: "The great corporations have them, the banks have them, all the organizations of business and of social and political activity have them."[42] It was the influence of public relations on the news that prompted Upton Sinclair, in 1919, to define journalism as "a business in the practice of presenting the news of the day in the interest of economic privilege."[43]

Today, with stepped-up public relations pressure from scientists and their institutions, journalists' negative attitudes have extended to science as well as business interests. Science journalists complain of the endless stream of public relations professionals: "I get calls from Doctor Knowledge, the world's leading authority on X disease or Y technology who is also president of

141

Z society." They refer to "pesky PR types" or "the flacks" who follow press releases with endless phone calls. "Any story that you might want to plant with me will, upon receipt of the phone call, have as much chance of making the pages of the *L. A. Times* as a run-over dog."[44] One reporter describes the "law of public relations lunches: The quality of news you get is inversely proportional to the quality of the lunch."

Resentment of public relations is evident in the National Association of Science Writers (NASW), the professional society of science journalists. Reflecting employment realities in the field, nearly half of NASW members are public relations writers working in universities or industrial firms. Though many are ex-science reporters and still freelance for science magazines, they are only allowed to be "associate members," unable to hold office. Not surprisingly, they resent their "second-class citizenship" and joke about the "caste system" and "separate toilets." But debates about their status have become increasingly intense as more and more journalists take public relations jobs.

Journalists' suspicion of public relations is by no means limited to the information coming from industry: "They're all grinding the same axe, from breakthrough university to wonder pharmaceuticals to the National Institute of Nearly Cured Diseases." Reporters still tend to trust scientists as sources, however, contrasting them with politicians: "When you talk to a scientist, you're talking to a fellow who is usually going to give you straight facts. His word is his bond. . . . When you talk to a politician, he is not worried about accuracy or truth; he usually reacts verbally and will say anything that comes to mind. He'll stretch the truth." "We would rather talk to scientists than to politicians because we know that they are getting at the truth." The fact that journalists resent manipulation by public relations officers and over-eager scientists does not diminish that influence—especially when sensitivity to manipulation is dulled by persistent faith in science as the ultimate, authoritative source of objective information. What is more difficult for journalists to accept is the reticence of scientists and their frequent efforts to avoid reporters. A part of public relations, after all, is an increasing effort to withhold sensitive information or to otherwise exercise communication controls over the news conveyed to the public.

9

HOW SCIENTISTS CONTROL THE NEWS

One day I talked to a geneticist who had agreed to hold a seminar on his research at the annual meeting of the National Association of Science Writers. He was worried and defensive. His scientific colleagues had warned him that he would be bombarded with value-laden questions about the potential applications of his work, its likely effects on industry-university relations, and its moral and ethical implications. In fact, the questions the science writers asked were limited to the technical issues he himself had raised; to his relief, the audience was simply struggling to understand the science. I asked him why he had been so concerned. "Well," he replied, "we scientists are working in a competitive and precarious environment. We work in a conspiratorial world." Science, he said, "is a never-never land of extraordinary risk . . . publicity is suspect. Whenever we say anything in public we are concerned about our image. We're afraid of what our colleagues will say."[1]

Despite the generally friendly tone and positive, even promotional, images that characterize much science and technology reporting, scientists complain of sensational-

144

ism and oversimplification. While they want their work to be covered in the press, they are constantly concerned about how it is covered, and this concern has led scientists and institutions not only to promote science through public relations, but also to control journalists' access to information.

Blaming the Messenger

Seeking a press that airs their views and supports their goals, scientists feel betrayed when their positions are challenged or distorted. They defensively interpret critical reports about science or technology as evidence of an antiscience or antiestablishment bias. In 1980, for example, Philip Handler, then president of the National Academy of Sciences, wrote with reference to the coverage of environmental disputes that "antiscience attitudes perniciously infiltrate the news media."[2] George A. Keyworth II, President Reagan's science adviser, asserted that reporters who cover science and technology deliberately distort facts: "the press is trying to tear down America."[3]

A scientist from UCLA wrote in a Dow Chemical publication, "The press often seems intent on showing how modern technology, including the chemical industry and nuclear power plants, is poisoning America." A chemical industry spokesman picked up on the image. "If there is any poisoning of America going on, it is not chemicals that are the culprit . . . it is the media, which all too often seem intent on burying us in piles of purple prose—a sort of verbal poison. In the process, journalists have helped to create crises where none exists (the cancer epidemic), have blown out of proportion legitimate stories (Three Mile Island) and avidly hunted for crises to come (acid rain). Now they seem bent on doing all this with the waste chemical issue."[4]

Even the most tempered and factual reporting can provoke a defensive response. When Harold Schmeck of the *New York Times* warned readers not to hope for immediate miracles from interferon research (see Chapter 1), scientists complained that such expressions of doubt by the press would affect their research funding: "It is the public's will and support that make further progress possible."[5]

Scientists complain about inaccuracy in the reporting of science and technology, telling horror stories of being misquoted, misinterpreted, or maligned. But when questioned further, they admit that inaccuracy is mainly a problem not of getting facts wrong but of omitting qualifications or details necessary to place information in a proper perspective. The nature of contemporary science journalism—the constraints of space and of time, the pressures for simplification, even the quasi-religious faith in the capacity of science to provide definitive solutions—and, sometimes, the ignorance of the reporter, are usually to blame for such omissions. Scientists, however, often suggest an explanation that ignores these complexities: journalists, they contend, simply care little about the truth. Arnold Relman, former editor of the *New England Journal of Medicine (NEJM),* put this bluntly: "The press, the media in general, are much more interested in the story, the news, than in the facts."[6]

Whatever the reasons for omissions and inaccuracies, scientists and public officials worry about the possible harm from media coverage of scientific and technological events. After the *Challenger* explosion a NASA official declared that the news media had pressured the agency to jeopardize flight safety in the space shuttle program.[7] In the aftermath of the artificial heart implantation at Humana Hospital, leading medical scientists suggested that the extensive reporting of the events could be detrimental to medical science and was less than a public service. Joseph Boyle, president of the AMA, referred to the publicity as a "Roman Circus" in which the desire to make "the deadline for evening news" could influence medical decisions. In the interest of scientific objectivity, he advocated less public reporting until the results were quietly and dispassionately reviewed by the scientific community.[8]

Following the initial publicity, even Humana's physicians became irritated with the press. Concerned about disruption (they found journalists hiding in laundry carts to get forbidden photographs) and reluctant to reveal the growing internal differences of opinion about the wisdom of the program, the hospital administration dismantled the media center and stopped the daily press briefings.

The ambivalence so many scientists feel toward the press stems less from their personal experience than from certain characteristics of their own profession in the context of growing social and economic pressures. The tradition of science places a high value on open communication. Although secrecy has also been part of the competitive culture of scientists—who are, after all, often engaged in priority disputes—the sharing of data is considered both a moral imperative and a pragmatic need. For secrecy is believed to be damaging to science, an obstacle to creativity, to the cumulative work necessary for progress, and to the peer review necessary to maintain the quality and integrity of scientific work.[9] Furthermore, scientists have often taken action to protect the public's right to know by opposing security restrictions, loyalty oaths, trade secrecy, and other measures that would restrict communication.

In recent years, however, the scientific enterprise has been undergoing significant changes that affect scientists' norms of communication, in particular their concerns about the press.[10] Science today has come to depend on sophisticated technology that is both extremely expensive and susceptible to regulation. The growing costs of science are changing the nature of both the scientific profession and its relation to the public, as criteria of social or commercial merit external to science are brought to bear on the funding and control of research. Increasingly, scientists are seeking support for large-scale projects from industry or directly from congressional appropriations, bypassing traditional agency sources (e.g., the National Science Foundation), which rely on scientific peer review. As competition for these corporate and government funds expands, many scientists are convinced that the traditional system of professional evaluation is not sufficient to sustain their work, that both research support and the ability to maintain autonomy will be affected by their public image.[11]

Changes in the relation of science to the public also reflect the growing economic, social, and policy importance of the knowledge generated by research. Control over scientific knowledge is integrally linked to power over public affairs. Thus people demand technical information, especially in controversial policy areas. For example, concern about the health effects of toxic wastes, environmental carcinogens, and chemicals in the workplace has been reflected in "right to know" legislation, in the demands by citizens for research data, and in greater use of the Freedom of Information Act. And as scientific information is perceived as a political resource, scientists themselves become involved as experts and advocates in environmental and other policy disputes. Their widely publicized participation inevitably corrodes the public image of their neutrality and disinterestedness, and this brings public pressure to bear on the process of science itself.[12]

Perhaps most important, the public image of science has changed in response to disclosures of misconduct, especially when episodes of data falsification have implications for scientific understanding of disease and, thus, for patient care. While scientists continue to claim that misconduct is rare, each new disclosure creates a public relations problem for the scientific community.

While scientists see public communication of scientific information as necessary and desirable, they are also aware that it extends their accountability beyond the scientific community. Once information enters the arena of public discourse, it becomes a visible public affair that is open to external investigation and regulation. Vulnerable to such external pressures and concerned about threats to their professional sovereignty, scientists have begun to seek more control of the public discourse on science— to influence the images of science in the press.

Strategies of Control

The scientific community has a well-articulated set of professional norms that govern relations among scientists. However, unlike physicians and other licensed professionals with codes of ethics and standards of confidentiality, they share few norms to guide their relations with the public. Defining their work as an autonomous enterprise, scientists are ill-equipped to deal with the external pressures represented by the media.

In an effort to mitigate the problem, scientific journals have published guidelines on how to respond to journalists and how to avoid misrepresentation. These guidelines are usually defensive. An article in the *NEJM* suggests that scientists use the public relations office of their universities as a clearinghouse. It warns scientists to be aware of reporters' motives: "Your response to an innocent-sounding question about your study of schizophrenia may be linked in tomorrow's newspaper to a murder trial." The article also warns *Journal* readers to avoid interviews about their research prior to publication and to be extremely cautious in what they say: "There are many instances in which a researcher has been led innocently to the slaughter," and "Never even whisper to a reporter anything you would not care to see in screaming headlines." It suggests that scientists who are to be interviewed first do a dry run with a public relations officer and also that they tape the interview so as to have an exact record. "If you feel trapped, obfuscate: it will get cut if it is too technical."[13]

Scientists attempt to control science news in part by discouraging their colleagues from "going public." In her book *The Visible Scientists,* Rae Goodell observed that those scientists who become visible to the media "are typically outsiders, sometimes even outcasts among established scientists, . . . seen by their colleagues almost as a pollution in the scientific community—sometimes irritating, sometimes hazardous. [They] are breaking old rules of protocol in the scientific profession, questioning the old ethic, defying the old standards of conduct."[14] Those who have the confidence to violate these norms are usually scientists with academic tenure and established reputations.[15]

Scientists' inclination to avoid reporters is encouraged by the editorial policies of a number of professional journals: the *Physical Review Letters, Archives of General Psychiatry,* the *NEJM, Science,* and several other journals will not consider an article whose content has been published in the popular press. These journals are highly prestigious; publishing in them affects a scientist's career. Thus their policies serve as effective constraints on scientific communication—often in research areas that are of broad public interest. Such policies have been a source of continuing and acerbic discussion, often focusing on the "Ingelfinger rule" guiding the publication policies of the *New England Journal of Medicine.*

In 1968 Franz J. Ingelfinger, then editor of the *NEJM,* decided that he would not publish a scientific article if the details of the article and its supporting data had been previously reported in another journal or in the press. He believed that the *NEJM* was more than simply archival; it should also be newsworthy. "The original article has an appeal quite different from that of the comprehensive survey. . . . The reader is more involved, his appetite is less dulled by the fact that his cerebral exposure to the news is direct, not through the dialyzing membrane."[16]

Ingelfinger's successor, Arnold Relman, perpetuated the rule. While Ingelfinger developed the policy because he wanted the *Journal* to be newsworthy, Relman emphasized his responsibility to maintain the reliability of scientific information through the system of peer review. He argued that the public interest is not well served by disregarding this system, for journalists could raise hopes or fears on the basis of false or unreliable information. In addition, Relman argued, prior disclosure places a burden on physicians, who should have the opportunity to read about research in an authoritative source before being besieged by patients clutching a newspaper article.[17]

Implementing the rule is problematic, given the normal practice of prepublication presentation of research at scientific meetings. Thus, Relman conceded that press coverage of papers presented at meetings could not preclude publication. However, he advised scientists not to grant interviews on the details of their conference papers, since their work at this stage is usually preliminary, incomplete, and may never warrant scientific publication.

Journalists are appalled by the Ingelfinger rule, arguing that it violates the public's right to know. They cite examples of delayed journal publication that had postponed appropriate public health measures. For example, a newspaper story about early laboratory research on the effects of "smoking" on beagles prevented publication of the findings of this cancer research, because scientific journals refused to publish the results on the grounds of prior disclosure in the press. Indeed, due to such concerns some journals, such as the *Journal of the American Medical Association (JAMA)* have refused to adopt the rule. Relman himself made exceptions in the *NEJM*. Early findings of research on toxic shock syndrome and AIDS, for example, were disclosed without jeopardizing the chance of publication because early public knowledge of medical information on these issues was urgent in light of the implications for public health.

Interpretations of urgency, however, can vary. In April 1994, an ABC program broke an embargo on a study to be published by the *New England Journal of Medicine*. The study had found, contrary to popular wisdom, that there was no evidence that vitamins protected against heart disease or cancer. While this was hardly a matter of urgency, the network officials argued that an embargo should not hold on an issue of such importance. Breaking the embargo in this case seemed principally to reflect the competitive quest for breaking news of wide popular interest.[18]

Ironically, the very newsworthiness of *NEJM* and *Science* (which, after all, are mailed to reporters before they are sent to subscribers) creates a struggle over the control of scientific

151

communication. Some scientists try to control press coverage by refusing interviews unless they can review and correct the copy prior to publication. Reporters, fearing censorship by vested interests, are reluctant to show their articles to sources, though they often confirm the accuracy of details with them. Science writer Victor Cohn suggests that "scientists are to reporters what rats are to scientists. Would scientists allow their subjects to check the interpretation of their behavior?"[19] Another writer, Earl Ubell, expanded on this characterization at a meeting of medical researchers: "As a reporter, I am neither proscience nor anti-science. To me as a reporter, the scientist is just like a rat is to you. I am looking at you through my microscope and trying to describe you. . . . What scientists and doctors do, not what you say about what you do, is ultimately what ought to be getting reported."[20]

During the Reagan and Bush administrations, federal and state governments introduced many restrictions on the flow of political information to the press, by imposing a system of prior censorship on government employees who wish to write about their work, and by reducing the number of press briefings. The government at this time also limited press access through changes in the Freedom of Information Act, budget constraints on the information-gathering activities of public agencies, and increased classification of federally funded research, even in areas that were not directly related to national security.

Washington, according to reporters, still "leaked like a sieve," but journalists who covered controversial issues complained of the growing difficulty of obtaining information on sensitive topics. A study of journalists' access to government sources during the 1980s found that the policies of federal agencies had become much more restrictive during that decade. Denied access to individual scientists, journalists were receiving only sanitized information that reflected official views.[21] These barriers have been gradually lifted in the more open climate of the Clinton administration.

Accidents or crises commonly generate efforts to exclude the press. The sensational and inaccurate reporting of the Three Mile Island accident, reflecting and then exacerbating the confusion that characterized this crisis, led to government and utility proposals to restrict media access to such events until official technical reports were available.[22] In April 1986, the U.S. Department of Energy and the Nuclear Regulatory Commission responded to the Chernobyl accident in the Soviet Union by placing gag orders on their own employees and contractors, including scientists working in national laboratories. They were to avoid the press or to limit their statements to background material. Despite their criticism of the Soviet Union for withholding details of the accident, these agencies themselves tried to limit public information. According to one memorandum, contractors and employees were not to make comparisons between U.S. and Russian reactors. Another voiced "a strong determination on the part of the U.S. government to speak with one voice" on matters relating to Chernobyl. All requests for information were to be channeled to three designated individuals.[23]

Scientists have also tried to impose restrictions on the media's access to sensitive events. During the controversy over recombinant DNA, research biologists excluded reporters from the 1975 meeting at the Asilomar Conference Center on potential risks of this research. They eventually allowed sixteen reporters to attend, but set boundaries on the discussion. It was to focus on technical questions of risk, avoiding philosophical issues of creating life, political issues of research funding, and social issues of the implications of genetic engineering. An embargo was placed on all stories until the conference was over.[24]

The possibility of patents, especially in biotechnology, has increased incentives for secrecy. Biotechnology research is expected to yield important agricultural, pharmacological, and diagnostic products. University research in these potentially profitable areas is often supported by interested industries in exchange for rights to license, produce, and market the results of the research. Industry has long provided support for university research, especially in engineering and chemistry departments.[25] But collaborations have significantly expanded since the early 1980s. Concerns about trade secrecy in these arrangements restrict the publication practices of scientists and, especially, their media comnmunications.

Scientists have also proposed restricting media access to information about ongoing medical research. At a 1975 conference on medicine and the media, medical researchers talked approvingly about establishing "truth squads" and "a central clearinghouse" that would verify scientific information before it was released to the press.[26] Physicians are concerned that press reports can have a harmful effect not only on medical practice but on their patients' behavior, as well. For example, in 1983 the *New England Journal of Medicine* published the preliminary results of research on a new therapy using hyperbaric oxygen for relief of the symptoms of multiple sclerosis. The scientific report was cautious, announcing that "preliminary results" suggested a "positive but transient effect" that warranted further study, and warning that the therapy could not generally be recommended. The press summaries of the report were also cautious in tone and included these qualifications. Nevertheless, hospitals were still bombarded with requests from multiple sclerosis victims and their families. And special private clinics opened storefront operations to administer oxygen.[27]

AIDS research is marked by frequent tensions between patients who are eager for reports about the latest therapies and scientists who want to restrict the release of information until the completion of controlled experimental studies. In 1989 a private group of AIDS researchers tried to impose an embargo on media reports of their trial of a new experimental drug, Compound Q. They feared that early publicity would end the study, so they promised some journalists access to data on the condition that they would not publish until the study ended; then those reporters would have exclusive rights. But the embargo did not work. As one reporter described the incident, concealing an AIDS study is like "hiding a rhinocerous in a cable car."[28]

Press reports about new scientific advances raise the hopes of desperate people. Dr. Louis Lasagna talks about the "awful impact" of *Reader's Digest* on the practice of medicine as patients come to their doctors' offices brandishing the latest copy and demanding the latest cure.[29] Dr. Larry Lamb, a medical writer whose syndicated column appears in 500 daily newspapers, receives about 20,000 hopeful letters from readers every month, including some from physicians asking for information to convey to their patients.[30]

When scientific research bears directly on health, it is often difficult to judge when best to release information to the press. How much evidence is necessary? How certain must the evidence be? Barry R. Bloom, chairman of the Department of Microbiology at the Albert Einstein College of Medicine, has argued that data should not be published until carefully evaluated in terms of "the totality of available evidence." Bloom believes that "until data are interpreted and validated, until the experimental design and significance are reviewed, and until all currently available data on the incidence . . . in exposed human populations can be integrated, the rush to the press is simply mindless if not unethical."[31] Taken literally, however, such constraints on releasing information to the public go beyond reasonable reticence; the press would wait indefinitely for medically related science news.

Concerns about controlling information are not limited to the medical profession. Richard Fiske, a geologist who directed the Museum of Natural History of the Smithsonian Institution, wants working scientists isolated from the media. Speaking at a professional society meeting, he described the chaotic reporting of a 1982 volcanic crisis at Guadalupe, an event marked by uncertainty, confusion, and scientific disagreement about the nature of risk. Reporters had talked directly to a number of scientists, each of whom offered a somewhat different interpretation of what was going on. Thus the media discovered and reported the disagreements within the scientific community. Fiske's response was to suggest that in the future all press communications be made through a professional intermediary or a public relations officer, who would assure that only "depersonalized" and "consistent" information be released. His proposal met with enthusiastic applause from his scientific colleagues.[32]

A similar proposal was brought to the Panel on Public Affairs of the American Physical Society. A physicist, concerned about public attitudes toward nuclear power and hoping to reduce "media distortion," wanted scientific societies to certify those scientists who were technically competent to speak to reporters.[33] Another scientist proposed the creation of a new profession of "certified public scientists" to make independent technical evaluations of scientific disputes and to be responsible for their public communication in those cases.[34]

These are paternalistic proposals; they are easily rationalized and some are reasonable, intended, for example, to reduce unwarranted public fear. But they place scientists in the inappropriate role of public guardians. Like most efforts to censor or limit information, they are also difficult to contain; for professional or political control over news can foster rumor, uninformed speculation, and sometimes exaggerated fears.[35]

Official accounts, selected and structured to emphasize the positive aspects of an issue, have sometimes caused costly delays in bringing a problem to public attention. For example, for years the press relied on official accounts about nuclear power, underreporting the concerns expressed by those outside the establishment. Simply acting as a conduit to transmit the views of their sources, journalists provided optimistic and positive views of a technology "too cheap to meter" that would solve the agricultural and energy problems of the world. Just before the accident at Three Mile Island, the local press covered the development of this facility by using only the information provided by utility officials, who consistently emphasized "good news."[36]

In the reporting of environmental problems, efforts by scientists or public officials to control communication have had high social costs, delaying public recognition of serious problems and preventing appropriate responses. The reluctance of scientists to talk directly to journalists was, according to Frank Graham, Jr., responsible for the long delay in public awareness of the pesticide problem: "In the face of man's massive intervention in the functioning of the natural world, the scientific establishment simply filed the ominous facts and kept mum. . . . They sneered at such techniques as popularization, and recoiled in indignation from the suggestion that they cooperate with the mass media to put across the story that should have been told."[37]

The press took over two years to report on an incident of food contamination in Michigan when a fire retardant chemical, PBB, was accidentally mixed with cattle feed in 1973. Local reporters, who relied on state agencies as sources of expertise, were simply reassured that the problem was contained. The national press did not give significant attention to the problem until it became an issue in the 1976 election campaign.

And the media was kept in the dark about the falsification of data in the breast cancer studies because those scientists and public officials who knew about the problem failed to report it for several years. Once the information became public knowledge in 1994 distraught patients and physicians felt, not surprisingly, betrayed.

These continuing efforts of scientists and scientific institutions to manage the media—through both public relations and communication controls—reflect their continued ambivalence about the press. Scientists today see improved press coverage as a means of fulfilling their obligation to bring science to the public. Selling science through the media is a way to attract support from legislators, corporate leaders, and foundation executives. But while seeking media coverage, scientists have also carried over values from a time when their disciplines were less accountable and more isolated from public affairs. Scientists worry about the corruptive influence of media visibility and about distortion of information, inaccuracy, and sensationalism. Thus, relations between science and the press remain fragile and strained. The question, however, is no longer *whether* science will be covered in the press, but *how* it will be conveyed. While the two cultures remain in tension, they are also inextricably bound.

10

THE HIGH COST
OF HYPE

On January 28, 1986, a long-standing and comfortable partnership between NASA and the press was shattered when the space shuttle *Challenger* exploded seconds after lift-off, killing all aboard.[1] The press reaction to the explosion was one of grief, disillusionment, and rage. For many longtime science journalists the event was a personal tragedy. "Those people were me," wrote a Houston reporter. "The shining star of technology for 30 years has dimmed." The *Miami Herald* compared the "countdown to disaster" to a "Greek tragedy, peppered with portents of the doom to come." The *New York Times* wrote of its disillusionment with an agency that "has symbolized all that is best in American technology . . . computerized, at the cutting edge of technology, sophisticated in its public relations strategy, squeaky-clean in its integrity."[2]

The space program had been important to the development of science journalism as a profession. The many months at Cape Canaveral had brought together journalists interested in science and technology, attracting new writers to the field. For three decades they had covered the space program as an awesome and pioneering venture, a source of national prestige. The first space shuttle in 1981 assumed symbolic dimensions in the popular press as an affirmation of American faith in science and technology, a "sweet vindication of American know-how." The press reports of the early space launches incorporated images that were for years to mark the media coverage of dramatic advances in science and technology. But then the *Challenger* accident became a watershed event, creating a sense of skepticism among journalists that was to influence their coverage not only of space technology, but also of other scientific and technical programs in subsequent years.

Fascinated with space technology, reporters had simply accepted what NASA fed them, reproducing the agency's assertions, promoting the prepackaged information they received, and rarely questioning the premises of the program, the competence of the scientists, or the safety of the operation. After the accident journalists felt betrayed. *Newsweek* announced that "the news media and NASA, wedded by mutual interest from the earliest days of the space program, are in the midst of a messy divorce." Having suddenly lost faith in the veracity of NASA, some newspapers even engaged in electronic war games, using high-technology interception antennas and experimental laser cameras to get stories that NASA wanted to conceal about the recovery of the shuttle.[3] Journalists were filled with self-incrimination, accusing themselves of accepting "spoon-fed news," of "treating the shuttle like a running photo opportunity," of letting readers down. More than any other event of that time, the *Challenger* accident aroused press and public awareness of the importance of probing, skeptical, and critical science journalism.

The *Challenger* accident was only one of a series of technological debacles during the 1980s that called attention to the cost of media hype. NASA experienced other failures with its much promoted megaprojects, including the $2 billion Hubble telescope and the $1 billion *Mars Orbiter*. The repair of the Hubble telescope in 1993 was celebrated but did not resurrect NASA's prestige; reporters observed the air of melancholy, the dispiriting sense of unfulfilled promise hovering over the space agency.

Other widely publicized projects have also backfired. Celebrated in 1987 as the "Colossus of Colliders" and "The Biggest Machine Ever," the Superconducting/Supercollider became, in 1993, a media metaphor for extravagant technologies. A Japanese accelerator used to kill malignant cells was called "the supercollider of cancer treatments, sapping money from smaller, but potentially more valuable projects." The debates over the Supercollider became a way for journalists to call attention to the inefficiency of megascience and the difficulties scientists have in managing enormous projects.

The same theme recurred in the 1994 coverage of the data falsification in a large-scale cancer therapy research program involving 89 hospitals. As the latest in a line of disclosures of fraud, this incident provoked a serious crisis of confidence in the self-regulating community of science. The revelation of the Cold War radiation experiments furthered the growing sense of betrayal. Following up on this event, reporters not only probed government archives at Department of Energy facilities, but also looked for similar scandals in old medical records at their local universities. Scientists these days are portrayed in less than awe-inspiring terms.

Science writers, in effect, are brokers, framing social reality for their readers and shaping the public consciousness about science-related events.[4] Their selection of news about science and technology sets the agenda for public policy. Their presentation of science news lays the foundation for personal attitudes and public actions. They are often our only source of information about the scientific and technical choices that significantly affect our work, our health, and our lives.

We pay for science and technology and bear their social costs. Public understanding of their social implications, their technical justifications, and their political and economic foundations is in the interest of an informed and involved citizenry. It is also critical to the future of science and technology themselves. The cost of public naïveté regarding science and the nature of scientific evidence has been apparent in many controversies—over the value of animal experimentation, appropriate precautions to prevent the spread of AIDS, and the teaching of evolution in the schools.

The media can play an important role in enhancing public understanding, but they have frequently failed to do so. There are many examples of brilliant science reporting, written with analytic clarity, critical insight, and provocative style, but too often science in the press is more a subject for consumption than for public scrutiny, more a source of entertainment than of information.[5] Too often science is presented as an arcane activity outside and above the sphere of normal human understanding, and therefore beyond our control. Too often the coverage is promotional and uncritical, encouraging apathy, a sense of impotence, and the ubiquitous tendency to defer to expertise. Focusing on individual accomplishments and dramatic or controversial events, journalists convey little about the sociology of science, the structure of scientific institutions, or the daily routines of research. We read about the results of research and the stories of success, but not about the process, the dead ends, the wrong turns. Who discovered what is more newsworthy than what was discovered or how. Thus science in the press becomes a form of sport, a "race" between scientists in different disciplines or between competitive nations, a rush to the "publication finish line."

There is little in this type of reporting to help readers understand the nature of scientific evidence and the difference between science and unverified opinion. Thus, they are ill-prepared to deal with scientific information even when it directly affects their interests. Many people, for example, are more confused than illuminated by reports of scientific findings about diet and cancer, or the genetic basis of breast and colon cancer. The persistent fear of catching AIDS by donating blood, despite scientific evidence to the contrary, is an important case in point.

In their promotional reporting of science and technology, writers still convey their fervent conviction that new technology will create a better world. But the message is increasingly polarized—we read of either promising applications or perilous effects, of triumphant progress or tragic risks. Impending breakthroughs are reported with zeal, technological failures with alarm. But the long-term political, economic, and social consequences of technological choices are seldom explored. Celebrating the development of each new medical technology, journalists failed to question the impact on health care costs until this became a central issue on the national policy agenda. As media oversell turned to the "genetic revolution" and journalists welcomed new gene discoveries and anticipated wonder cures, they gave little attention to the ethical and social implications of predicting genetic disease—the problems of privacy and the possibilities of genetic discrimination—until scientists themselves began to worry about such issues.

This book has suggested that many of the characteristics of science and technology reporting reflect the nature of the relationship between journalists and their sources. Concerned about their legitimacy in the political arena and anxious to receive support for their work, scientists are sensitive to their image in the press. Hoping to shape that image, they are becoming adept at packaging information for journalists. Like advocates in any field, they are prone to overestimate the benefit of their work and minimize its risks.

For their part, journalists, especially those with limited experience in science reporting, are vulnerable to manipulation by their sources of information. They are concerned about balance and objectivity and accept the ideology of science as a neutral authority, an objective judge of truth. Some science writers are in awe of scientists; others are intimidated. But most are bewildered by the complexity of technical issues. The difficulty of evaluating a complex and uncertain subject converges with the day-to-day constraints of the journalistic profession to reinforce the tendency to rely uncritically on scientific expertise. While political writers often go well beyond press briefings to probe the stories behind the news, science reporters tend to rely on authoritative researchers, press conferences, and professional journals. The result? Many journalists have adopted the mindset or "frame" of scientists, interpreting science in terms defined by their sources, even when those sources clearly display a special bias.

Thus while art, theater, music, and literature are routinely subjected to criticism, science and technology are almost always spared—until outrageous incidents occur. While political writers aim to analyze and criticize, science writers seek to elucidate and explain. There are few outlets for journalists who would serve as critical commentators on, or probing investigators of, science and technology. Rare are the Walter Lippmanns or I. F. Stones of science who write regularly in the press.[6] Unaggressive in their reporting and reliant on official sources, science journalists present a narrow range of coverage. Many journalists are, in effect, retailing science and technology rather than investigating them, identifying with their sources rather than challenging them.

I have suggested throughout this book that the pattern is beginning to change in the 1990s; there is more critical, more skeptical science reporting. But as critical reporting increases, so do the tensions between science and journalists; for these tensions reflect fundamental differences in the norms and values of these two professions.

To begin with, they often differ in their judgments about what is news. To scientists, research results become reliable and therefore newsworthy through replication and endorsement by professional colleagues. Prior to publication in reputable journals, scientific papers are carefully evaluated and approved through the system of peer review. This system of establishing reliability has been critical to the structure of science, especially to the process of scientific communication. For scientists, then, research findings are tentative, undigested, provisional—and therefore not newsworthy—until certified by peers to fit into the existing framework of knowledge. For journalists, on the other hand, certified and established ideas are "old news"—of far less interest than fresh and dramatic, though possibly tentative, research. Seeking to entertain as well as to inform, they are attracted to nonroutine, nonconventional, and even aberrant events.[7] This difference between scientists and journalists becomes a source of contention when overzealous researchers, as in the cold fusion case, seek press coverage of "hot" research prior to the time-consuming process of peer review. Such hubris is especially contentious in the 1990s when many scientists fear that changes in science—the increased dependence on industry support and the incidents of misconduct—are undermining the peer review and refereeing processes.

Journalists seeking credible views on controversial issues, often rely on the opinions of scientists who have become public figures. Nobel Prize winners, because of their general prestige, are frequently cited in fields well outside their specialized expertise. Scientists suspect such use of unverified opinion. Arnold Relman, former editor of the *New England Journal of Medicine,* expressed the scientists' view: "If a [politician] makes a statement of what the policy of his government is or what he thinks or what he is going to vote, that's news. . . . News of a new development in science is coupled with evidence. Opinion is not important, it's evidence. Opinion is cheap and can be misleading in science, but opinion in politics or public affairs is another matter."[8]

Some journalistic practices conflict with scientific expectations about appropriate styles of communication. While both groups are committed to communicating truth, journalists must often omit the careful documentation and precautionary qualifications that scientists feel are necessary to present their work

165

accurately. While scientists are socialized to qualify their findings, journalists may see qualifications as protective coloration. Furthermore, readability in the eyes of the journalist may be oversimplification to the scientist. Indeed, many accusations of inaccuracy are traceable to reporters' efforts to present complex material in a readable and appealing style.

Journalistic conventions intended to enhance audience appeal may also violate scientific norms. To make abstract technical decisions more concrete, science writers often examine the personal choices of their technical informants ("Would you live at Love Canal?"), undermining the idea that technical decisions are based on depersonalized evidence. To create a human interest angle, journalists also personalize research, writing on the foibles and features of individuals. But the focus on individual accomplishments and the presentation of scientists as stars contradicts communal norms, which favor a collective image of science as an objective and disinterested profession. Similarly, to convince their editors about the newsworthiness of science and technology, journalists emphasize the uniqueness of individual events (the "first" discovery, the "breakthrough," the "biggest" collider). Although many scientists actively contribute to the breakthrough syndrome, ideally they prefer to emphasize continuity and the cumulative nature of research.

Journalists relate their stories to immediate and familiar social and economic contexts. In the 1990s they have questioned the high cost of megaprojects and looked to the short-term applications of science. Scientists see the need of long-term research strategies and are often threatened by the media's focus on immediate economic and social goals.

The journalistic preoccupation with conflict and aberration, intended to attract the reader's interest, is a further source of strain. In covering disputes journalists tend to create polarities: technologies are either risky or they are safe. The quest for simplicity, drama, and brevity precludes the complex, nuanced positions that scientists prefer. But the polarized presentation of technical disputes also reflects journalists' norms of objectivity— their belief that verity can be established by balancing conflicting claims. This approach further contributes to strain, for to a scientist objectivity is based on the understanding that claims must be verified by empirical means—hardly by balancing opposing views.

Differences in the use of language add to the tension. The language of science is intended to be precise and instrumental. Scientists communicate for a purpose—to indicate regularities and aggregate patterns, to provide technical data. In contrast, journalistic language has literary roots. Journalists will choose words for their richness of reference, their suggestiveness, their graphic appeal. They are likely to prefer a "toxic dump" to a "waste disposal facility."

In any discourse, language is organized to address the background and assumptions of the anticipated audience.[9] Scientists direct their professional communication to an audience trained in their discipline. They take for granted that their readers share certain assumptions and therefore will assimilate the information conveyed in predictable ways.[10] Journalists, on the other hand, write for diverse readers whose interpretations will vary with their interests, objectives, and technical sophistication.[11] Thus, while scientists talk of aggregate data, reporters write of the immediate concerns of their readers: Should I use saccharin? Drink coffee? Take estrogen? "Will I be harmed?"[12]

Often words that have a special meaning in a scientific context will be interpreted differently by the lay reader. For example, the word "epidemic" has both technical and general connotations. Scientists use the word "epidemic" to describe a cluster of incidents greater than the normal background level of cases. If the background level is zero, then six cases are technically an epidemic. To the public or the journalist, an epidemic implies thousands of cases, a rampantly spreading disease.

Confusion over the definition of "evidence" occurs among scientists as well as in the press, often confounding the discourse of risk disputes. Biostatisticians use the word "evidence" as a statistical concept. But for biomedical researchers the critical experiment may also be defined as evidence. Most lay people accept as credible evidence anecdotal information or individual cases. So, too, do journalists. Such differences frequently lead to misunderstanding. In reports about the health effects of exposure to toxic chemicals at Love Canal, for example, scientists and journalists held different assumptions about the definition of credible evidence concerning the validity of animal tests, the neighborhood's habitability, and the adequacy of containment of the chemicals. Thus, when scientists described the health effects of dioxin with a cryptic "no evidence" (meaning no statistically significant evidence), journalists interpreted their response as an effort to cover up the problem, since they knew of individual cases.

Perhaps the most important source of strain between scientists and journalists lies in their differing views about the appropriate role of the press. Scientists often talk about the press as a conduit or pipeline, responsible simply for transmitting science to the public in a way that can be easily understood. They expect to control this flow of information to the public as they do within their own domain. Confusing their special interests with general questions about the responsibility of the press, they are reluctant to tolerate independent analysis of the limits or flaws of science. They assume that the purpose of science journalism is to convey a positive image that will promote science; they see the press as a means of furthering scientific goals.

This view of journalism is reflected in scientists' complaints about the press and its effects on public attitudes. Scientists tend to attribute negative public attitudes about science and technology to problems of media communication. They blame journalists, who, they believe, distort the flow of information from scientists to the public. Problems of scientific communication could, however, as easily be attributed to the *sources* of information—to suppression of facts, to manipulation of conclusions, or to overeager, promotional public relations.[13]

Many science journalists see their mission as one of recording "official history"—of elucidating, amplifying, and even eulogizing science. But others, in recent years, are questioning their role as "self-appointed trumpets" for science and technology. Reacting to technological failures, to the economic implications of large and costly scientific endeavors, and to changes taking place within scientific fields that are increasingly tied to corporate interests, many journalists are beginning to suspect promotional hype. They are questioning the authority of science, raising probing questions in their interviews with scientists: Who pays? Who is responsible? What's in it for the public? What are the stakes?

"Gee whiz," "cosmic breakthrough" articles continue to appear in the press coverage of science and technology, but many journalists today are more critical. They believe that "it is not enough for us to report the new discoveries or gadgetries; we must delve deeper into their effects on people and public policy." One reporter I interviewed wants "to take some of the awesomeness out of science." Another hopes "to create a better-informed citizenry able to deal with problems." These are among the goals frequently expressed by journalists today.

Professional organizations in the field of science journalism constantly work to improve the standards of reporting. The National Association of Science Writers (NASW) was founded in 1934 by a dozen veteran science writers[14] to "foster the dissemination of accurate scientific knowledge by the press of the nation in cooperation with scientific organizations and individual scientists."[15] Struggling to convince their editors that science was news, the NASW founders hoped that professionalizing their specialty would enhance its visibility, recognition, and prestige.

When press coverage of science and technology began to expand after World War II, so too did the NASW membership:

there were 113 members in 1950, 413 in 1960, 830 in 1970, 1200 in 1986, and 1830 in 1993. The NASW newsletter, *Science Writers,* disseminates professional "gossip" and special articles of interest to the advancement of a growing profession.[16] In 1960 NASW spawned an independent nonprofit organization, the Council for the Advancement of Science Writing (CASW), a group of writers, editors, educators, and scientists who raise funds to support the development of science journalism programs and annual briefings (the "New Horizons in Science" program) in which distinguished scientists talk to journalists about current scientific advances.

Scientists have also initiated efforts to assure that timely and reliable information reaches the press. The Scientists' Institute for Public Information (SIPI) includes a Media Resource Service (MRS) with a computer file of over 25,000 scientists and engineers who have agreed to answer queries from reporters. It receives about 75 telephone inquiries a week from journalists seeking reliable sources on a wide range of subjects. When a crisis such as an earthquake or oil spill occurs, hundreds of reporters call in. If the subject of inquiry is a disputed one, MRS routes journalists to several scientists selected to represent a spectrum of opinion. The organization also brings together scientists and journalists to discuss controversial issues such as animal experimentation or the disposal of toxic wastes. The purpose is to enhance both the technical sophistication of journalists and the political sophistication of scientists potentially involved in disputes.[17]

Scientists and journalists are negotiating the public meaning of science and technology. This terrain, today more than ever, is contested as journalists increasingly—and appropriately—probe issues of scientific responsibility and accountability, questioning the ideologies and social priorities that guide science policy decisions. The tensions between these two communities are inevitable; indeed, maintaining their differences is essential if each community is to fulfill its unique social role.

If the popular press is to act as a watchdog over major social and political institutions, if it is to mediate between science and the public and facilitate public discourse about crucial policy issues, both scientists and journalists must accept and come to terms with an uneasy and often adversarial relationship. Scientists must restrain the promotional tendencies that lead to controls on information or to oversell, and they must open their doors to more vigorous investigation.

Journalists, for their part, must try to convey understanding as well as information. It is not enough merely to react to scientific events, translating and elucidating them for popular consumption. To comprehend science or technology, readers need to know its context: the social, political, and economic implications of scientific activities, the nature of evidence underlying decisions, and the limits—as well as the power—of science as applied to human affairs.

NOTES

PREFACE

1. These are the subject of another book: Dorothy Nelkin and Susan Lindee, *The DNA Mystique: The Gene as a Cultural Icon* (New York: W. H. Freeman, 1995).

2. Studies of the relative importance of various media include Serena Wade and Wilbur Schramm, "The Mass Media as Sources of Public Affairs, Science and Health Knowledge," *Public Opinion Quarterly* 33, summer 1969, pp. 197–209; Lawrence W. Lichty, "The News Media," *Wilson Quarterly* 6, 1982, pp. 49–57; and Evan Witt, "Here, There and Everywhere: Where Americans Get Their News," *Public Opinion* 6, August/September 1983, pp. 45–48.

3. This journal, edited by John Durant, is published at the British Science Museum in London.

CHAPTER 1

1. From Slosson's "Talks to Trustees," quoted in David J. Rhees, *A New Voice for Science: Science Service under Edwin E. Slosson, 1921–29,* Master's thesis, University of North Carolina, 1979.

2. Frank Carey, "A Quarter Century of Science Reporting," *Neiman Reports* 20, June 1966, pp. 7–8.

3. For a review of the history of the research developments, see Sandra Panem, *The Interferon Crusade* (Washington, D.C.: Brookings, 1984).

4. Harold Schmeck's controversial article appeared in the *New York Times* on May 27, 1980. The reply appeared on June 17, 1980.

5. Neal Lane, quoted in *Nature* 366, November 1993, p. 395.

6. Part of this increase reflects efforts by the National Science Foundation to enhance the public understanding of science. See Bruce Lewenstein, "Was there really a Science Boom?" *Science, Technology and Human Values,* spring 1987, pp. 29–41.

7. See discussion of scientists' criticism of the press in June Goodfield, *Reflections on Science and the Media* (Washington, D.C.: AAAS, 1981). See also Jay A. Winsten, "Science and the Media: The Boundaries of Truth," *Health Affairs* 4, spring 1985, pp. 5–23. Some excellent articles on the relationship between scientists and journalists appear in Sharon Friedman, Sharon Dunwoody, and Carol Rogers (eds.), *Scientists and Journalists* (New York: The Free Press, 1986).

8. William Burrows, "Science Meets the Press: Bad Chemistry," *Sciences,* April 1980, pp. 15–19.

9. In a study of Washington reporters, Stephen Hess asked them which papers they read regularly. He found that 89 percent read the *Washington Post,* 73 percent the *New York Times,* and 51 percent the *Wall Street Journal.* Hess also surveyed public officials. Of his sample of high federal officials, 90 percent read the *Post,* 45 percent the *Times,* and 62 percent the *Wall Street Journal.* Stephen Hess, *The Washington Reporters* (Washington, D.C.: Brookings, 1981).

10. The press also includes alternative newspapers that assume an explicit advocacy role and often focus on specific policy issues.

Suburban and community newspapers cover little science. Science news in special sections of scientific journals such as *Science* and the *New England Journal of Medicine* is directed to the scientific community, not the lay public, and is not included in this study.

11. Studies have suggested similarities in the way journalists from different countries construct their stories despite differences in the organization of the press. Sharon Dunwoody and H. Peters, "Mass Media Coverage of Technological and Environmental Risks: A Survey of Research in the United States and Germany," *Public Understanding of Science,* 1, April 1992, pp. 199–230.

12. Todd Gitlin, *The Whole World is Watching* (Berkeley: University of California Press, 1980), p. 7. See also Gaye Tuchman, *Making News* (New York: The Free Press, 1978).

13. See William Garson and Kathryn Lasch, *Evaluating the Welfare State* (New York: Academic Press, 1983).

14. Edward Lawless, *Technology and Social Shock* (New Brunswick: Rutgers University Press, 1977); Herbert Gans, *Deciding What's News* (New York: Vintage Books, 1980); Michael Schudson, *Discovering the News* (New York: Basic Books, 1978); and Warren Breed, *The Newspapermen: News and Society* (New York: Arno Press, 1980).

15. Kenneth Burke, "Literature as Equipment for Living," *The Philosophy of Literary Form* (Berkeley: University of California Press, 1973), p. 298.

16. George Lakoff and Mark Johnson, in *Metaphors We Live By* (Chicago: University of Chicago Press, 1980), suggest how metaphors define reality. See also Murray Edelman, *Political Language* (New York: Academic Press, 1977).

CHAPTER 2

1. *Nation,* January 16, 1902.

2. *Boston Globe,* October 15, 1993.

3. *Kansas City Times,* September 17, 1982.

4. *Time,* October 27, 1980.

5. *New York Times,* February 19, 1983.

6. *New York Times,* June 12, 1984.

7. *McCall's,* July 1964, pp. 38–40, 124.

8. *Science Digest* 55, February 1964, pp. 30–36.

9. *Vogue,* January 1978, p. 174.

10. *Family Health,* June 1978, p. 24.

11. *Newsweek,* October 24, 1983.

12. *New York Times,* October 11, 1983.

13. *Time,* October 29, 1979.

14. *New York Times,* November 30, 1993.

15. Edward O. Wilson, *On Human Nature* (Cambridge, Mass.: Harvard University Press, 1980).

16. For a comprehensive review of the criticism and the controversy, see Ullica Segerstrale, "Colleagues in Conflict: An In Vivo Analysis of the Sociobiology Controversy," *Biology and Philosophy* 1, 1986, pp. 53–87.

17. *Time,* August 1, 1977.

18. *Cosmopolitan,* March 1982.

19. Camille Benbow and Julian Stanley, "Sex Differences in Mathematical Reasoning: Fact or Artifact?" *Science* 210, December 12, 1980, pp. 1262–1264.

20. *New York Times,* December 7, 1980.

21. Pamela Weintraub, "The Brain: His and Hers," *Discover,* April 1981, pp. 15–20.

22. *Time,* December 15, 1980.

23. *Time,* January 20, 1992, p. 42.

24. *Business Week,* April 10, 1978.

25. *New York Times,* November 30, 1993.

26. *New York Times,* January 23, 1994.

27. See, for example, interviews in the *New York Times,* October 12, 1975, and *People Weekly,* November 19, 1975.

28. *Newsweek,* April 12, 1976.

29. *Science Digest,* March 1982.

30. *New York Times,* January 23, 1994.

31. *New York Times,* July 12, 1994.

32. *Boston Globe,* June 19, 1980.

33. *New York Times,* August 5, 1980.

34. *Christian Science Monitor,* March 10, 1982.

35. See, for example, the *Washington Post,* August 14, 1991.

36. Judith Swazey, Melissa S. Anderson, and Karen Seashore Louis, "Ethical Problems in Academic Research, *American Scientist,* 81, November–December 1993, pp. 542–53.

37. *Washington Post,* November 13, 1993.

38. See, for example, the *New York Times* May 5, 1989: April 1, 1991; and July 22, 1991.

39. See for example, Nicholas Wade, "Method and Madness," New *York Times Magazine,* December 26, 1993.

40. *Newsweek,* February 8, 1982.

41. For examples of the media view of industry-university relations, see Sheldon Krimsky, *Biotechnics and Society* (New York: Praeger, 1991), p. 71.

CHAPTER 3

1. *U.S. News and World Report,* December 28, 1981; January 4, 1982.

2. *New York Times,* January 3, 1994.

3. *Salt Lake Tribune,* March 1983.

4. *U.S. News and World Report,* September 15, 1980.

5. *New York Times,* January 9, 1994.

6. *Christian Science Monitor,* June 9, 1982.

7. *Columbus Evening Dispatch,* January 23, 1983.

8. See, for example, *New York Times,* January 23, 1994.

9. *Newsweek,* October 18, 1957.

10. *Newsweek,* August 9, 1982.

11. *New York Times,* September 18, 1983.

12. *Newsweek,* March 7, 1983, p. 67.

13. Kenneth Wilson, cited in *SIPIScope* 13, November/December 1985, p. 11.

14. *New York Times,* December 29, 1993.

15. *New York Times,* January 2, 1994.

16. These quotations are from the local newspapers that are catalogued by issue in *Newsbank* 1982 and 1983.

17. For coverage of the conference, see Sheldon Krimsky, *Genetic Alchemy* (Cambridge, Mass.: MIT Press, 1982) and Michael Rogers, *Biohazard* (New York: Knopf, 1977). For a review of news coverage, see Michael Altimore, "The Social Construction of a Scientific Controversy," *Science, Technology and Human Values,* fall 1982, pp. 24–31.

18. See, for example, the use of the language of miracles in a feature article, "The Miracle of Spliced Genes," *Newsweek,* March 17, 1980, and in *U.S. News and World Report,* December 12, 1979.

19. Liebe Calaviere, "New Strains of Life or Death?" *New York Times Magazine,* August 22, 1976.

20. See, for example, *Birmingham Alabama News,* May 25, 1983.

21. *Time,* January 17, 1994.

22. See, for example, the *New York Times,* January 30, 1994.

23. See Dorothy Nelkin and Laurence Tancredi, *Dangerous Diagnostics* (Chicago: University of Chicago Press, 2nd edition, 1994).

24. Renee Fox and Judith Swazey, *Spare Parts* (New York: Oxford University Press, 1992).

25. See, for example, *Life,* September 17, 1971.

26. Lawrence K. Altman, "After Barney Clark: Reflections of a Reporter on Unsolved Issues," in Margery Shaw (ed.), *After Barney Clark* (Austin: University of Texas Press, 1984), pp. 113–128.

27. *New York Times,* December 3, 1982; and *Philadelphia Inquirer,* December 5, 1982.

28. *Salt Lake Tribune,* December 2, 1982.

29. *Salt Lake Tribune,* December 3, 1982.

30. Nancy Sommers, Princeton, New Jersey, unpublished list of media reports and articles on ERT.

31. For example, see *San Diego Union,* December 13, 1964; *Pittsburgh Post Gazette,* November 10, 1964; and *Ladies' Home Journal,* January 1965.

32. Robert A. Wilson, "The Roles of Estrogen and Progesterone in Breast and Genital Cancer," *Journal of the American Medical Association,* 182, October 1962, pp. 327–331; and *Feminine Forever,* (New York: M. Evans, 1966).

33. AP report, April 14, 1965, quoting Francis P. Rhoads.

34. AP report, January 29, 1966.

35. *Look,* January 11, 1966.

36. Robert Kistner developed his ideas in a popular book entitled *The Pill: Fact and Fallacies* (New York: Delacorte Press, 1969) and in the *Ladies' Home Journal* and other women's magazines in 1969.

37. *McCall's,* October 1971.

38. *Vogue,* January 1974.

39. *New York Times,* December 5, 1975.

40. Letter from Stanley Sauerhaft, Hill and Knowlton, to William Davis, Ayerst Laboratories, December 17, 1976.

41. Elliot Valenstein, *Great and Desperate Cures* (New York: Basic Books, 1986) describes the press coverage of psychosurgery in the 1940s.

42. Waldemar Kaempffert, "Science in the News: Psychosurgery," *New York Times,* January 11, 1942.

43. See, for example, articles in *Time,* July 24, 1978; July 31, 1978; August 7, 1978; February 19, 1979; and in *Newsweek,* July 24, 1978.

44. See the discussion of these debates and their press coverage in Dorothy Nelkin and Chris Anne Raymond, "Tempest in a Test Tube," *The Sciences* 20, November 1980.

45. Some of these scenarios are from the *Newsletter of the Center for Biotechnology Policy and Ethics of Texas A&M University,* which reviewed the media coverage of human embryo cloning on January 1, 1994.

46. November 8, 1993.

47. October 27, 1993.

CHAPTER 4

1. Eleanor Singer and Phyllis Endreny, *Reporting on Risk,* (New York: Russell Sage Foundation, 1993). For analysis of cases of risk reporting, see Dorothy Nelkin, background paper for 20th Century Fund, *Science in the Streets* (New York: Priority Press, 1984).

2. Karl Duke, "Myths of nature: Culture and the Social Construction of Risk," *Journal of Social Issues* 48, 1992, 21–37.

3. J. E. Lovelock, R. J. Maggs, and R. J. Wade, "Halogenated Hydrocarbons in and over the Atlantic," *Nature* 241, January 19, 1973, pp. 194–196.

4. Lydia Dotto and Harold Schiff, *The Ozone War* (New York: Doubleday, 1978).

5. Mario J. Molina and F. Sherwood Rowland, "Stratospheric Sink for Chlorofluoromethanes: Chlorine Atom-Catalyzed Destruction of Ozone," *Nature* 249, June 28, 1974, pp. 810–812.

6. These and similar articles appeared throughout 1975. See Dotto and Schiff, op. cit., chap. 7.

7. Cited in Dotto and Schiff, op. cit., p. 251.

8. Ibid., p. 169.

9. See discussion in Paul Brodeur, "Annals of Chemistry," *New Yorker,* June 9, 1986, pp. 70–87.

10. *New York Times,* September 14, 1976; *Washington Post,* September 14, 1976.

11. Michael Brown and Katherine Lyon, "Holes in the Ozone Layer," in Dorothy Nelkin (ed.), *Controversy: Politics of Technical Decisions,* 3d ed. (Newbury Park, Calif.: Sage, 1992).

12. Lee Wilkins, "Between Facts and Values," *Public Understanding of Science* 2, January 1993.

13. *Business Week,* October 18, 1969.

14. *Newsweek,* March 12, 1979.

15. *Time,* August 26, 1974.

16. Daniel Greenberg, "Washington Report," *New England Journal of Medicine,* 300, March 22, 1979, pp. 687–688.

17. *New York Times,* February 28, 1979.

18. *New York Times,* March 3, 1979.

19. *Washington Post,* March 3, 1979. See also March 5 and 7, 1979.

20. See, for example, reports in *Newsweek* and *Time* during March and April 1977.

21. *Newsweek,* April 4, 1977. See also March 21, 1977.

22. *Time,* April 25, 1977.

23. See Adeline Levine, *Love Canal: Science, Politics and Rights* (Lexington, Mass.: D. C. Heath, 1982).

24. Phil Brown and Edwin Mikkelson, *No Safe Place* (Los Angeles: University of California Press, 1990).

25. *Time,* September 22, 1980.

26. *Los Angeles Times,* May 9, 1983.

27. *Milwaukee Journal,* June 19, 1983.

28. *Atlanta Journal,* June 4, 1983.

29. *Arkansas Gazette,* June 17, 1983.

30. *Time,* May 2, 1983.

31. *Washington Post,* January 22, 1983.

32. *New York Times,* February 23, 1993.

33. Sheldon Krimsky, *Biotechnology and Society* (New York: Praeger, 1990).

34. See, for example, *Barrons,* May 31, 1993, and the *New York Times Magazine,* November 21, 1993, p. 24.

35. I would like to acknowledge a course paper written by Anthony Panzica, graduate student at New York University, fall 1993.

CHAPTER 5

1. This term is from Peter Sandman, "Mass Media and Environmental Risk," *Risk,* July 1994.

2. Reported by Eliot Marshall, "Public Attitudes to Technological Progress," *Science* 205, July 20, 1979, pp. 281–285.

3. See data in National Science Board, *Science and Engineering Indicators—1989,* Washington, D.C.: US Government Printing Office 1989 (NSB 89–1).

4. For a review of the efforts to establish the impact of the media see Denis McQuail, "The Influence and Effects of Mass Media" in James Curran, M. Gurevich, and J. Woollacott (eds.), *Mass Communication and Society* (Beverly Hills, Calif.: Sage, 1979), chap. 3.

5. Charles Cooley, *Human Nature and the Social Order* (New York: Scribners, 1922). The work on communication by Cooley and his colleagues is reviewed in Daniel Czitrom, *Media and the American Mind* (Chapel Hill: University of North Carolina Press, 1982), chap. 4.

6. Walter Lippmann, quoted in the *Conservationist* 33, May/June 1979, p. 1.

7. Paul Lazarsfeld, Bernard Berelson, and Hazel Gaudet, *The People's Choice* (New York: Duell, Sloan and Pierce, 1944).

8. Raymond and Alice Bauer, "America, Mass Society and Mass Media," *Journal of Social Issues* 16, 1960, pp. 3–66.

9. Joseph Klapper, *The Effects of Mass Communication* (Glencoe, N.Y.: Free Press, 1960).

10. See discussion in Dorothy Nelkin and Susan Lindee, *The DNA Mystique: The Gene as a Cultural Icon* (New York: W. H. Freeman, 1995).

11. George Gerbner et al., "Scientists on the TV Screen," *Society,* May/June 1981, pp. 41–44. See also Marcel LaFollette, "Science on Television: Influences and Strategies," *Daedalus,* fall 1982, pp. 183–197; Jonathan Weiner, "Prime Time Science," *Science,* September 1980, pp. 6–13; and Nancy Signorielli, "Science and Television: Some Images and Viewer Conceptions of Social Reality," paper presented at the annual meeting of the American Academy of Arts and Sciences, Detroit, May 1983 (mimeo).

12. George Basalla, "Pop Science: The Depiction of Science in Popular Culture," in Gerald Holton and William Blanpied (eds.), *Science and Its Public* (Dordrecht: D. Reidel, 1976), p. 266.

13. Marcel LaFollette, *Authority, Promise and Expectation: The Image of Science and Scientists in American Popular Magazines, 1910–1955,* Ph.D. thesis, University of Indiana, 1979, pp. v, xiv. Material from her thesis appears in her book *Making Science Our Own* (Chicago: University of Chicago Press, 1990).

14. See Donald N. Let, "Four Wondrous Weeks of Science and Medicine in the Amazing, Incredible Supermarket Press" NASW *Newsletter,* January 1980.

15. For a review of this coverage, see Curtis MacDougal, *Superstition and the Press* (Buffalo, N.Y.: Prometheus Books, 1983).

16. John Burnham, *How Superstition Won and Science Lost* (New Brunswick: Rutgers University Press, 1987).

17. National Science Board, op.cit., p. 164.

18. Louis Harris Poll, commissioned by SIPI, reported in *SIPI-Scope* 20, spring 1993.

19. National Cancer Institute, Cancer Prevention Awareness Survey, NIH #84–26–77 (Washington, D.C.: Government Printing Office, 1984), pp. 58, 64.

20. Colin Seymour-Ure, *The Press, Politics and the Public* (London: Methuen, 1968); and James B. Lemert, *Does Communication*

Change Public Opinion After All? (Chicago: Nelson-Hall, 1981), pp. 44–45.

21. William Gamson and André Modigliani, "Media Discourse and Public Opinion on Nuclear Power," *American Journal of Sociology* 95, July 1989, pp. 1–37.

22. Oscar Handlin, "Science and Technology in Popular Culture," *Daedalus* 94, winter 1965, p. 156.

23. Gilbert Omenn, "Are Kids Afraid to Become Scientists?" *Science* 83, September 1983, p. 18.

24. Jon Miller, K. Prescott, and R. Pearson, *The Attitudes of the U.S. Public Towards Science and Technology* (Chicago: National Opinion Research Center, 1980). See also Daniel Yankelovitch, "Changing Public Attitudes to Science and the Quality of Life," *Science, Technology and Human Values* 39, spring 1982, pp. 23–29.

25. Susanna Hornig, "Reading Risk: Public Responses to Print Media Accounts of Technological Risk," *Public Understanding of Science* 2, April 1993, pp. 95–110.

26. *Newsweek,* March 26, 1990.

27. See, for example, *American Health,* October 1990.

28. Peter Kramer, *Listening to Prozac* (New York: Viking, 1993).

29. *Newsweek,* February 7, 1994.

30. Edward J. Robinson, "Analyzing the Impact of Science Reporting," *Journalism Quarterly* 40, 1963, pp. 306–314.

31. Ronald Troyer and Gerald Markle, *Cigarettes: The Battle over Smoking* (New Brunswick, N.J.: Rutgers University Press, 1983).

32. This "agenda setting" role has been widely documented. See M. E. McCombs and D. L. Shaw, "The Agenda Setting Function of the Mass Media," *Public Opinion Quarterly* 36, summer 1972,

pp. 176–187; Bernard C. Cohen, *The Press and Foreign Policy* (Princeton: Princeton University Press, 1963): Graham Murdock, "Mass Communication and the Construction of Meaning," in Nigel Armstead (ed.), *Reconstructing Social Psychology* (Harmondsworth, England: Penguin, 1974); and Gaye Tuchman, *Making News* (New York: The Free Press, 1978).

33. Allan Mazur and K. M. Waba, "The Mass Media at Love Canal and Three Mile Island," paper presented at the annual meeting of the American Sociological Association, Detroit, September 2, 1983. Mazur contends that the quantity of coverage of a controversial technology, regardless of the bias, creates negative attitudes. See Allan Mazur, *The Dynamics of Technical Controversy* (Washington, D.C.: Communications Press, 1981).

34. Gladys Lang and Kurt Lang, *The Battle for Public Opinion* (New York: Columbia University Press, 1983).

35. Harold Laski, *The American Democracy* (New York: Viking Press, 1984).

36. Adeline Levine, *Love Canal: Science, Politics and People* (Lexington, Mass.: D. C. Heath, 1982).

37. Sheldon Krimsky, *Genetic Alchemy* (Cambridge, Mass.: MIT Press, 1983).

38. Gerald Markle (ed.), *The Politics of Laetrile* (Beverly Hills, Calif.: Sage, 1982).

39. Joseph D. Cooper, "Cyclamate Sequel: Risks Dominating Benefits," *Medical Tribune,* February 26, 1970, p. 11.

40. Sandra Panem, *The Interferon Crusade* (Washington, D.C.: Brookings, 1984).

41. On influences on the direction of research, see Terry Shinn and Richard Whitley (eds.), *Expository Science* (Dordrecht: D. Reidel, 1985).

42. The role of the press in this area is documented in Steven Maynard-Moody, "The Fetal Research Dispute," in Dorothy Nelkin (ed.), *Controversy: The Politics of Technical Decisions,* 3d ed. (Newbury Park, Calif.: Sage, 1992).

43. Dorothy Nelkin and Judith P. Swazey, "Science and Social Control," in Ruth Macklin (ed.), *Research on Violence* (New York: Plenum, 1981).

CHAPTER 6

1. Quoted in Carolyn Hay, *A History of Science Writing in the United States,* Master's thesis, Northwestern University, 1970, p. 35.

2. There are few historical accounts of science journalism. See Hay, ibid.; David J. Rhees, *A New Voice for Science: Science Service under Edwin E. Slosson 1921–29,* Master's thesis, University of North Carolina, 1979; also Marcel LaFollette, *Making Science Our Own: Public Images of Science 1910–1955* (Chicago: University of Chicago Press, 1990). Frank Luther Matt in his classic history of magazines (*A History of American Magazines,* vols. 1–5, Cambridge, Mass.: Harvard Press, Belknap Press, 1968; originally published, 1930) has useful reviews of science journalism.

3. On the popularization of physics, see Daniel Kevles, *The Physicists* (New York: Knopf, 1978), chap. 2.

4. Hay, op. cit., p. 34.

5. Quoted in LaFollette, op. cit., p. 189.

6. Frederick Lewis Allen, *Only Yesterday: An Informal History of the Twenties* (1931, reprint ed. New York; Harper & Row, 1964), pp. 164–165.

7. Quoted in Ronald Tobey, *The American Ideology of National Science: 1919–1930* (Pittsburgh: University of Pittsburgh Press, 1971), p. 106.

8. Hay, op. cit.

9. Quoted in Kevles, op. cit., p. 171.

10. Rhees, op. cit., pp. 31–32, and Tobey, op. cit.

11. Science Service, "A Statement of Purpose," *Science News Letter,* August 11, 1928, p. 90.

12. From Slosson's "Talks to Trustees," quoted in Rhees, op. cit., pp. 38–39.

13. Edwin Slosson, *Chats on Science* (New York: Century, 1924), p. 169.

14. Edwin Slosson, "Democracy of Knowledge," in B. Brownell (ed.), *Preface to the Universe* (New York: Van Nostrand, 1929), p. 108.

15. William Laurence, cited in Spencer Weart, *Nuclear Fear* (Cambridge: Harvard University Press, 1989) chap. 5.

16. Waldemar Kaempffert, *New York Times,* September 22, 1919; cited in Weart, op. cit., chap. 1.

17. Quoted in obituary for Gobind Lal, NASW *Newsletter,* April 1982.

18. Remarking on the concern about objectivity, Herbert J. Gans, *Deciding What's News* (New York: Random House, 1979), p. 184, calls journalism "the strongest remaining bastion of logical positivism in America."

19. William L. Rivers, Wilbur Schramm, and Clifford Christian, *Responsibility in Mass Communication* (New York: Harper & Row, 1980), Appendix A.

20. Ibid., Appendix B.

21. Dan Schiller, *Objectivity and the News* (Philadelphia: University of Pennsylvania Press, 1981).

22. Ibid., pp. 80, 87.

23. Quoted ibid., p. 180.

24. Quoted in Michael Schudson, *Discovering the News* (New York: Basic Books, 1978), pp. 78–80.

25. Quoted in Schiller, op. cit., p. 194.

26. See discussion in Edward Jay Epstein, *News from Nowhere* (New York: Random House, 1973).

27. Hay, op. cit.

28. Gaye Tuchman, "Objectivity as Strategic Ritual," *American Journal of Sociology* 77, 1972, pp. 660–679, and *Making News* (New York: Free Press, 1978).

29. Todd Gitlin, *The Whole World is Watching* (Berkeley: University of California Press, 1980).

30. Harvey Molotch and Marilyn Lester, "News as Purposive Behavior," *American Sociological Review* 39, February 1974, pp. 101–112.

31. Personal interview with *Science* reporter Luther Carter.

32. NASW *Newsletter,* March 1966.

33. John Lear, "The Trouble with Science Writing," *Columbia Journalism Review* 9, 1972, p. 30.

34. David Perlmam, "Science and the Mass Media," *Daedalus* 103, summer 1974, p. 207.

35. This was also the time of considerable expansion of the advocacy (or political) press, which denied all pretense of objectivity. See Chris Anne Raymond, *Uncovering Ideology: Occupational Health in the Mainstream and Advocacy Press,* Ph.D. thesis, Cornell University, 1983.

36. Perlman sent me his files of all his bylined articles for three years, representing three periods in his career. Obviously, because journalists write about what is going on, no selection of years can be used as typical of a writer's interests. For example, in 1976–1977 Perlman and other science reporters were preoccupied with

genetic engineering. Other years nuclear power issues dominated their columns. I selected these three years to suggest the range and tone of reporting at different times.

37. Personal correspondence with Perlman, January 24, 1984.

38. Quoted in Scott DeGarmo, "An Editor Takes a Survey: Are Scientists Better Science Writers Than Non-Scientists?" NASW *Newsletter,* October 1981, pp. 16–18; see also Pierre C. Fraley, "The Education and Training of Science Writers," *Journalism Quarterly* 40, 1963, pp. 323–328.

39. Nicholas Wade, "Scientists and the Press: Cancer Scare Story That Wasn't," *Science* 174, November 1971, pp. 679–680.

40. For background on the education of science writers, see Hillier Krieghbaum, *Science and the Mass Media* (New York: New York University Press, 1967), chap. 6, and Lee Z. Johnson, "Status and Attitudes of Science Writers," *Journalism Quarterly* 34, spring 1957, pp. 247–251.

41. From testimony in 1974 to the House of Representatives Select Committee on Labor; cited in an editorial in *Columbia Journalism Review,* September/October 1977, pp. 8–9.

42. Bob Hall, "The Brown Lung Controversy," *Columbia Journalism Review,* March/April 1978, p. 33.

43. Wade Roberts, "Phosvel: A Tale of Missed Cues," *Columbia Journalism Review,* July/August 1977, p. 23.

44. Betty Medsger, "Asbestos: The California Story," *Columbia Journalism Review,* September/October 1977, pp. 41–47.

45. Quoted in Hall, op. cit., p. 35.

46. Gans, op. cit. See also W. C. Johnstone et al., *The News People* (Urbana: University of Illinois Press, 1976).

47. NASW *Newsletter,* August 1979, p. 4.

48. Rae Goodell, "How to Kill a Controversy: The Case of Recombinant DNA," in Sharon M. Freedman, S. Dunwoody, and Carol Rogers (eds.), *Scientists and Journalists* (New York: The Free Press, 1986), pp. 170–181.

49. David Perlman, self-portrait, in *Science in the Newspaper* (Washington, D.C.: AAAS, 1974), p. 8.

50. Ibid.

51. Victor McElheny, "Reporting of a Golden Age of Discovery", paper presented at the Conference on Expository Science, Paris, December 1–3, 1983, pp. 2–3 (mimeo).

52. Ed Edelson, "Science Writing Celebrates the Trivial," NASW *Newsletter,* December 1980.

53. These quotations and others without citation are from personal interviews, informal conversations and science writers, and their comments at meetings and press conferences.

Chapter 7

1. H. Schwartz, "AIDS in the Media" in *Science in the Streets* (New York: Twentieth Century Fund, 1984); and Dorothy Nelkin, "AIDS and the News Media," *Milbank Quarterly,* 69, 1991.

2. Randy Shilts, *And the Band Played On* (New York: St. Martins Press, 1987).

3. Dorothy Nelkin, David Willis, and Scott Parris, *A Disease of Society* (New York: Cambridge University Press, 1991).

4. For discussion of these constraints as they affect general reporting, see Herbert J. Gans, *Deciding What's News* (New York: Random House, 1979) and Edward Jay Epstein, *News from Nowhere* (New York: Random House, 1973). For the application to science reporting, see articles in Sharon Friedman, Sharon Dunwoody, and Carol Rogers (eds.), *Scientists and Journalists* (New York: The Free Press, 1986).

5. Time constraints may also affect reporting in different parts of the country. If a press conference is held at two o'clock and the deadline is five, writers will have little time to flesh out the story with background details. West Coast writers attending a scientific meeting on the East Coast will have the benefit of the time change. The timing of events with respect to deadlines can affect whether a topic gets covered and the thoroughness of the report.

6. Communications and Medical Research Symposium Proceedings, University of Pennsylvania, October 17, 1964, p. 32.

7. *SIPIScope* 20, fall 1992.

8. Sharon Dunwoody, "The Science Writing Inner Club," *Science, Technology and Human Values* 3, winter 1980, pp. 14–22.

9. David Perlman, *American Scientist,* May 1988.

10. Leo Bogart, "Editorial Ideals, Editorial Illusions," in Anthony Smith (ed.), *Newspapers and Democracy* (Cambridge, Mass.: MIT Press, 1980), pp. 247–278.

11. Kenneth Johnson, "Dimensions of Judgment of Science News Stories," *Journalism Quarterly* 40, 1963, pp. 315–322.

12. David Perlman, NASW *Newsletter,* December 1978.

13. Larry Sabato, *The Rise of Political Consultants* (New York: Basic Books, 1981).

14. *New York Times,* April 14, 1994.

15. Similar avoidance of controversy is sometimes evident among the editors of scientific journals. For example, two NIH investigators had conducted a study of professional misconduct in science and found a high frequency of practices departing from accepted standards. Though the article had been reviewed by peers as a public service, no journal editor was willing to publish their results. See correspondence from the files of Walter Stewart and Ned Feder of the National Institutes of Health during 1983 and

1984 covering their paper on "Professional Practice among Biomedical Scientists."

16. For a review of these efforts see Geoffrey Sea, "The Radiation Story No One Would Touch," *Columbia Journalism Review,* March/April, 1994, pp. 37–40.

17. Oliver S. Moore III, quoted in Bernice Kanner, "Scientific Experiments: Too Many Books?" *New York Magazine,* October 29, 1984, p. 16.

18. John Shelton Lawrence and Bernard Timbers, "News and Mythic Selectivity," *Journal of American Culture* 2, summer 1979, pp. 321–330.

19. Ben H. Bagdikian, "Conglomeration, Concentration and the Media," *Journal of Communication* 30, spring 1980, pp. 59–64.

20. Peter Dreier and Steve Weinberg, "Interlocking Directorates," *Columbia Journalism Review,* November/December 1979, pp. 51–68.

21. American Newspaper Publishers Association, *Facts about Newspapers* (Washington, D.C.: ANPA, 1982).

22. See discussion in John Westergard, "Power, Class and the Media," in James Curran, M. Gurevitch, and J. Woollacot, (eds.), *Mass Communication and Society* (Beverly Hills, Calif.: Sage, 1979), pp. 95–115.

23. Quoted in Robert Cirino, *Don't Blame the People* (New York: Vintage Books, 1972), p. 93.

24. Alan Lupo, quoted in Dreier and Weinberg, op. cit.

25. Dwight Jensen, "The Loneliness of the Environmental Reporter," *Columbia Journalism Review,* January/February 1977, pp. 41–42.

26. Sharon Dunwoody, "Community Structure and Media Coverage of Risk," *Risk,* July 1994.

27. This term was coined by Steven Hungerford and James B. Lemert, "Covering the Environment: New Afghanistanism?" *Journalism Quarterly* 50, autumn 1973, pp. 475–481, 508.

28. Jeffrey Kirsch, "On a Strategy for Using the Electronic Media to Improve Public Understanding of Science and Technology," *Science, Technology and Human Values* 4, spring 1979, pp. 52–58.

29. J. Lukasieweics, "The Ignorance Explosion," *Transactions (New York Academy of Science), 1972, pp. 373–390. Also, Derek de Solla, Price, Science Since Babylon* (New Haven: Yale University Press, 1961).

30. For reviews of accuracy in the press reports of complex scientific information see Sharon Dunwoody, "A Question of Accuracy," *IEEE Transactions on Professional Communication* PC-25, December 1982, pp. 196–199; James W. Tankard, Jr., and Michael Ryan, "Untangling the Numbers: Journalists Can Cope with Complex Research," *Newspaper Research Journal* 3, July 1982, pp. 61–69; and William Witt, "Effects of Quantification on Scientific Writing," *Journal of Communication* 1, winter 1976, pp. 67–69.

31. Dunwoody, op. cit.

32. See Sharon Friedman, "Environmental Reporting: Problem Child of the Media," *Environment* 25, December 1983, pp. 24–29; and Philip Tichenor, "Teaching and the Journalism of Uncertainty," *Journal of Environmental Education* 10, spring 1979, pp. 5–8.

33. For a review of these technical uncertainties about the health effects of exposure to chemicals, see introduction and technical appendix of Dorothy Nelkin and Michael S. Brown, *Workers at Risk* (Chicago: University of Chicago Press, 1984).

34. Stephen Hilgartner, "The Diet Cancer Debate," in Dorothy Nelkin (ed.), *Controversy: Politics of Technical Decisions,* 3d ed. (Newbury Park, Calif.: Sage, 1992).

35. Anthony Smith, *Goodly Gutenberg: The Newspaper Revolution of the 1980s* (New York: Oxford University Press, 1980), p. 188.

36. David Blum, "A Kafkaesque Tale of Health Faddists Eating Cockroaches and Journalists Eating Crow," *Wall Street Journal,* September 28, 1981.

37. Daniel Boorstin, *The Image* (New York: Harper & Row, 1964).

38. Quoted in William Rivers, *The Adversaries: Politics and the Press* (Boston: Beacon Press, 1970), p. 49.

39. Personal communication with Fred Golden of *Time,* November 1983.

40. Quoted in Sharon Friedman, *Changes in Science Writing Since 1965 and Their Relation to Shifting Public Attitudes Toward Science,* Master's thesis, Pennsylvania State University, 1974, p. 66.

41. From discussions during a meeting of the Twentieth Century Fund Task Force on Risk and the Media, November 1983.

42. David B. Sachsman, "Public Relations Influence on Coverage of Environment in San Francisco Area," *Journalism Quarterly* 53, spring 1976, pp. 54–60.

43. Leon V. Sigal, *Reporters and Officials* (Lexington, Mass.: D. C. Heath, 1973).

44. Mary Marlino, *Reporting on PCBs in the Hudson,* Master's thesis, Cornell University, 1984.

45. Nancy Hicks, statement in discussion during a symposium on medicine and the media, University of Rochester Medical Center, October 9–10, 1975, *Proceedings,* p. 133.

CHAPTER 8

1. F. Sherwood Rowland, "President's Lecture: The Need for Scientific Communication with the Public." *Science* 260, 11 June 1993, pp. 1571–6.

2. Quoted in the *New York Times,* March 16, 1985.

3. For discussion of the norms and communication practices of science, see Robert Merton, *Sociology of Science* (New York: University of Chicago Press, 1973), and Bernard Barber and Walter Hirsch (eds.), *The Sociology of Science* (New York: The Free Press, 1962).

4. For a history of this period see Bruce Lewenstein, "The Meaning of Public Understanding of Science," *Public Understanding of Science* 1, 1992, pp. 45–68.

5. Quoted in Hillier Krieghbaum, "American Newspaper Reporting of News," *Kansas State Bulletin,* August 15, 1941, p. 20.

6. Ronald Tobey, *The American Ideology of National Science 1919–1930* (Pittsburgh: University of Pittsburgh Press, 1971).

7. NASW *Newsletter,* June 1, 1954, p. 10.

8. George Kistiakowsky, speech to the National Association of Science Writers, quoted in "Our Twenty-Fifth Anniversary," NASW *Newsletter* 7, December 1959, pp. 11–12.

9. Pat McGrady, public relations officer of the American Medical Association, in a 1967 review quoted in Carolyn Hay, *A History of Science Writing in the United States,* Master's thesis, Northwestern University, 1970, p. 254.

10. Jean Rostand, "Popularization of Science," *Science* 131, May 1960, p. 1491.

11. Quoted in Gerard Piel, "Democracy and the Intellect," in Miriam Balaban (ed.), *Science Information Transfer: The Editor's Role* (Amsterdam: Elsevier, 1978).

12. Robert H. Grant and Kenneth D. Fisher, "Scientists and Science Writers: Concerns and Proposed Solutions," *FASEB Proceedings* 39, 1971, pp. 816–826.

13. *New York Times,* April 25, 1986.

14. Edward Feigenbaum and Pamela McCorduck, *The Fifth Generation* (Reading, Mass.: Addison-Wesley, 1983), p. 8.

15. See Wilson interview in *SIPIScope* 13, November/December 1985, p. 11.

16. Arthur Caplan, "So How Are We Doing?" paper presented at the annual meeting of the American Society of News Editors, April 10, 1985.

17. Phil Gunby, "Media-Abetted Liver Transplants Raise Questions of Equity and Decency," *Journal of the American Medical Association* 249, April 1983, pp. 1973–1982.

18. *Science Writer,* summer 1988, 1–3.

19. *Courier Journal,* November 27, 1984.

20. *Courier Journal,* March 3, 1985.

21. Jay Winsten, "Science and the Media: The Boundaries of Truth," *Health Affairs* 4, spring 1985, pp. 5–23.

22. Dan Rather, "CBS News," cited in Bruce Lewenstein. "Cold Fusion and Hot History," *Osiris,* 2nd series, 1992, 7: 135–163.

23. This event has been the subject of several popular books, including J. R. Huizinga, *Cold Fusion: The Scientific Fiasco of the Century* (Rochester: University of Rochester Press, 1992) and Gary Taubes, *The Short Life and Weird Times of Cold Fusion* (New York: Random House, 1993). See also Thomas Gieryn, "The Ballad of Pons and Fleischmann," in Ernan McMullin (ed.), *The Social Dimension of Science* (Notre Dame: University of Notre Dame Press, 1992), pp. 217–43; and Dale Sullivan, "Exclusionary Epideictic," *Science, Technology and Human Values,* vol. 19, summer 1994, pp. 283–306.

24. *Boston Globe,* October 7, 1993.

25. Robert S. Cowen, "Garbage Under Glass: What Are Scientists Dishing Out?" *Technology Review* 81, November 1979, pp. 10–11.

26. Many industrial scientists are willing to play this role. Studies suggest that while corporate scientists may face conflicts between their expectations of scientific autonomy and the pragmatic demands of bureaucratic loyalty, they tend nonetheless to adopt managerial values, which include concerns about enhancing corporate image. This loyalty guides their response to the press. For studies of industrial science see William Kornhauser, *Scientists in Industry* (Berkeley: University of California Press, 1962); Barney Glaser, *Organizational Scientists* (Indianapolis: Bobbs Merrill, 1964); R. Ritti, *The Engineer in the Industrial Corporation* (New York: Columbia University Press, 1971); and Edwin A. Layton, *The Revolt of the Engineers* (Cleveland: Case Western Reserve Press, 1971).

27. Richard Tucker, lecture to Clinical Institute of Toxicology, 1983 (mimeo).

28. Cited in Stephen Hilgartner, Richard Bell, and Rory O'Connor, *Nukespeak* (San Francisco: Sierra Club Books, 1982), p. 77.

29. Ibid., p. 79.

30. Michael D. Tabris, "Surviving a State of Siege," *The Chemist,* May 1983, pp. 5, 22.

31. Richard Tucker, op. cit.

32. Information excerpted from their public relations material. In 1982, the scientists' lectures were reported in Atlanta, San Bernardino, Houston, Phoenix, Oklahoma City, and Charleston.

33. These phrases are from public relations brochures.

34. *Science Writer,* spring 1991, 1–4.

35. For examples of such "breakthroughs" see Jim Sibbison, "Covering Medial Breakthroughs," *Columbia Journalism Review,* July/August, 1988, pp. 36–39.

36. Francesca Lunzer, "Medicine by Press Agentry," *Forbes,* September 24, 1984, pp. 199–202.

37. See *Science* 224, June 1984, pp. 1392–1398. Critiques are reviewed in *Medical World News,* August 27, 1984, pp. 38–46 and *Science* 225, August 1984, pp. 705–706. See also the *Washington Post,* June 23, 1984.

38. I. Stanton, "Public Relations and the Press," in Gary Wagner (ed.), *Publicity Forum* (New York: R. Weiner, 1977), p. 31.

39. "Declarations of Principles," published in William L. Rivers, Wilbur Schramm, and Clifford Christian, *Responsibility in Mass Communications* (New York: Harper & Row, 1980), Appendix E.

40. NASW *Newsletter,* passim, 1979.

41. David Zimmerman, "How Consumer Education, Public Relations Style, Still Translates as 'Press Agentry'," NASW *Newsletter,* April 1979.

42. Quoted in Michael Schudson, *Discovering the News* (New York: Basic Books, 1978), p. 139.

43. Quoted in Dan Schiller, *Objectivity and the News* (Philadelphia: University of Pennsylvania Press, 1981), p. 188.

44. These statements are from letters and discussion in the NASW *Newsletter,* from which innumerable other examples could also be drawn. See, for instance, Mark Bloom, "The Editor's Letter," NASW *Newsletter,* April 1979, and letters to the editor, passim.

CHAPTER 9

1. Personal discussion with a visiting speaker at the NASW New Horizons Meeting, Blacksburg, Virginia, November 1983.

2. Philip Handler, "Public Doubts about Science" (editorial), *Science* 108, June 6, 1980.

3. George Keyworth was interviewed by Fred Jerome in *SIPIScope,* February 1985.

4. "The Poisoning of America or . . . the 'Poisoning' of America," in *The Point Is . . .* (Dow publication), October 7, 1980.

5. Harold Schmeck's article appeared in the *New York Times* on May 27, 1980; the reply appeared on June 17, 1980.

6. Arnold Relman comment in "Medicine in the Media: Panel Discussion," *P & S* 2, April 2, 1982, p. 16.

7. *New York Times,* March 16, 1986.

8. Quoted in Lawrence K. Altman, "Publicity and Medicine," *New York Times,* January 1, 1985.

9. For discussion of the norms of open communication see "The Normative Structure of Science," in Robert K. Merton, *The Sociology of Science* (Chicago: Chicago University Press, 1973), pp. 267–278. Also see Sissela Bok, "Secrecy and Openness in Science," *Science, Technology and Human Values* 7, winter 1982, pp. 32–41.

10. For discussion of the changing pressures on science and their effect on open communication see Dorothy Nelkin, *Science as Intellectual Property* (New York: Macmillan, 1984).

11. Sharon Dunwoody and Michael Ryan, "Scientific Barriers to the Popularization of Science—The Mass Media," *Journal of Communication* 35, winter 1985, pp. 26–42.

12. See discussion in Dorothy Nelkin (ed.), *Controversy: Politics of Technical Decisions,* 3d ed. (Newbury Park, Calif.: Sage, 1992).

13. Martin Bander, "The Scientist and the News Media," *New England Journal of Medicine* 308, May 1983, pp. 1170–1173. See also Neal E. Miller, "The Scientist's Responsibility for Public Information: A Guide to Effective Communication with the Media," *Society for Neuroscience* (Bethesda, Maryland: 1978).

14. Rae Goodell, *The Visible Scientists* (Boston: Little Brown, 1977), p. 61.

15. Sharon Dunwoody and Byron T. Scott, "Scientists as Mass Media Sources," *Journalism Quarterly* 59, spring 1982, pp. 52–59; Luc Boltanski and Pascale Maldidier, "Carrière Scientifique, Morale Scientifique et Vulgarisation," *Information Science and Society* 9 (3), 1970, pp. 99–118.

16. Franz J. Ingelfinger, "Medical Literature: The Campus without Tumult," *Science* 169, August 1970, p. 733.

17. Relman has written and lectured extensively on his policy. See *New England Journal of Medicine* 303, December 1980, pp. 1527–1528; *Bryn Mawr Alumni Bulletin,* fall 1981, pp. 2–5; NASW *Newsletter,* November 1979, pp. 9–10; and "Special Report on Medicine and the Media," *P & S* 2, April 1982, p. 16.

18. Reported in the *New York Times,* April 14, 1994.

19. Quoted in June Goodfield, *Reflections on Science and the Media* (Washington, D.C.: American Association for the Advancement of Science, 1981), p. 94.

20. Symposium on medicine and the media, University of Rochester Medical Center, October 9–10, 1975, *Proceedings,* p. 182.

21. Carol Rogers, "Reporters Access to Agency Sources," *Science Writers,* spring 1991, 1–4.

22. See discussion in President's Commission on the Accident at Three Mile Island, *Report of the Public Right to Information Task Force* (Washington, D.C.: Government Printing Office, 1979).

23. *New York Times,* May 22, 1986.

24. Michael Altimore, "The Social Construction of a Scientific Controversy: Comments on Press Coverage of the Recombinant DNA Debate," *Science, Technology and Human Values* 7, fall 1982, pp. 24–31.

25. Dorothy Nelkin, *Science as Intellectual Property* (New York: Macmillan, 1984).

26. Symposium on medicine and the media, University of Rochester Medical Center, October 9–10, 1975, *Proceedings*, p. 34.

27. *Philadelphia Inquirer*, September 25, 1983.

28. *Science Writer*, fall 1989, 1–3.

29. Louis Lasagna, symposium on medicine and the media, University of Rochester Medical Center, October 9–10, 1975, *Proceedings*, p. 7.

30. Ibid., p. 58.

31. Barry R. Bloom, "News about Carcinogens: What's Fit to Print," *Hastings Center Report*, August 1979, pp. 5–7.

32. NASW *Newsletter*, April 1, 1982.

33. Reported in William Colglazier, Jr., and Michael Rice, "Media Coverage of Complex Technological Issues," in Dorothy Zinberg (ed.), *Uncertain Power* (New York: Pergamon, 1983), p. 113.

34. G. Bugliarello, "A Technological Magistrature," *Bulletin of the Atomic Scientists* 1, 1978, pp. 34–37; and J. C. White, "Reflections and Speculations on the Revelation of Molecular Genetic Research," *Annals of the New York Academy of Sciences* 265, 1976, p. 173.

35. Kerry Smith et al., "Can Public Information Programs Affect Risk Perceptions?" *Journal of Policy Analysis and Management*, 9, 1990, pp. 41–59.

36. Sharon M. Friedman, "Blueprint for Breakdown: Three Mile Island and the Media before the Accident," *Journal of Communication* 31, spring 1981, pp. 116–128.

37. Frank Graham, Jr., *Since Silent Spring* (Boston: Houghton Mifflin, 1970), pp. 165–166.

CHAPTER 10

1. This section on the *Challenger* accident was developed with the help of Susan Lindee when she was a Cornell University doctoral student. She is now a Professor at the University of Pennsylvania.

2. *Houston Chronicle,* February 1, 1986; *Miami Herald,* February 23, 1986; and *New York Times,* February 5, 1986.

3. *New York Times,* March 20, 1986.

4. For a theoretical perspective on the hegemonic role of the press see Stuart Hall, "Culture, the Media and the Ideological Effect," in James Curran, M. Gurevitch, and J. Woollacott (eds.), *Mass Communication and Society* (Beverly Hills, Calif.: Sage, 1979), chap. 13.

5. See discussion in Robert Young, "Science as Culture," *Quarto,* December 1979, p. 7; and Langdon Winner, "Mythinformation," in Paul T. Durbin (ed.), *Research in Philosophy and Technology,* vol. 7 (Greenwich, Conn.: JAI Press, 1984), pp. 287–304.

6. There are, of course, notable exceptions, including Daniel Greenberg, who publishes a newsletter entitled *Government and Science Report* and sometimes writes a science policy column for the *Washington Post.* David Zimmerman publishes a newsletter called *Probe.* The *New York Times* has hired several reporters from *Science* who have been writing interpretive and critical articles. The news and comments section of *Science* also publishes critical commentary; however, *Science* is a specialized publication read mainly by scientists and science policy professionals.

7. L. E. Trachtman, "The Public Understanding of Science Effort," *Science, Technology and Human Values,* vol. 6, summer 1981, pp. 10–15.

8. Arnold Relman, "Special Report on Medicine and the Media," *P & S,* April 1982, p. 22.

9. See Erving Goffman, "Felicity's Condition," *American Journal of Sociology* 89, July 1983, pp. 1–53.

10. J. C. Pocock, "Ritual Language and Power," *Politics, Language and Time* (London: Methuen, 1970).

11. Richard Whitley, in Terry Shinn and Richard Whitley (eds.), *Expository Science* (Dordrecht: D. Reidel, 1985), points out that the many different nonscientific groups that constitute the audience for scientific information seek such information for specific purposes and assimilate it accordingly.

12. See discussion in Harold Sharlin, *EDB: A Case Study in the Communication of Health Risk,* (Washington, D.C.: Environmental Protection Agency, 1985).

13. See, for example, David Rubin, "What the President's Commission Learned about the Media," in T. Moss and D. Sills (eds.), *The Three Mile Island Accident* (New York: New York Academy of Sciences, 1981), pp. 95–106.

14. Howard Blakeslee (Associated Press), Ferry Colton (Associated Press), Watson Davis (Science Service), David Dietz (Scripps Howard), Victor Henderson (*Philadelphia Inquirer*), Thomas Henry (*Washington Star*), Waldemar Kaempffert (*New York Times*), Gobind Behari Lal (*Hearst*), William Laurence (*New York Times*), John O'Neill (*New York Herald Tribune*), Robert B. Potter (*American Weekly*), and Allen Shoenfield (*Detroit News*).

15. The charter is reproduced in Carolyn D. Hay, *A History of Science Writing in the United States,* Master's thesis, Northwestern University, 1970.

16. Another professional newsletter, *Sciphers,* appeared in 1979, published by the Science Writing Educators Groups at the University of Missouri.

17. A similar organization, the Center for Health Communication, has been formed at the Harvard School of Public Health to clarify and interpret health information for journalists. The communications office of the American Association for the Advancement of Science plays a similar role.

INDEX

ABOUT THE AUTHOR

Dorothy Nelkin holds a University Professorship at New York University, teaching in the Department of Sociology and the School of Law. She was formerly on the faculty of Cornell University. Her research is in the area of science, technology, and society, where she focuses on the relationship between science and the public as expressed in disputes, public perceptions of science, and institutional responses to scientific information.

She is a member of the National Academy of Sciences' Institute of Medicine, a fellow and former Director of the American Association for Advancement of Science, a fellow of the Hastings Center, and a member of the Ethical, Legal and Social Implications Working Group of the NIH Human Genome Project. She has been a Guggenheim Fellow, a former President of the Society for Social Studies of Science, and winner of their Bernal Prize.

Her books include: *Controversy: Politics of Technical Decisions*; *Science as Intellectual Property*; *The Creation Controversy*; *Workers at Risk*; *Dangerous Diagnostics: The Social Power of Biological Information* (with L. Tancredi); and *The DNA Mystique: The Gene as a Cultural Icon* (with S. Lindee).